A Building Revolution: How Ecology and Health Concerns Are Transforming Construction

DAVID MALIN ROODMAN

AND

NICHOLAS LENSSEN

Jane A. Peterson, *Editor*

WORLDWATCH PAPER 124
March 1995

THE WORLDWATCH INSTITUTE is an independent, nonprofit environmental research organization based in Washington, D.C. Its mission is to foster a sustainable society—in which human needs are met in ways that do not threaten the health of the natural environment or future generations. To this end, the Institute conducts interdisciplinary research on emerging global issues, the results of which are published and disseminated to decisionmakers and the media.

FINANCIAL SUPPORT is provided by The Nathan Cummings Foundation, the Energy Foundation, the Geraldine R. Dodge Foundation, The George Gund Foundation, W. Alton Jones Foundation, John D. and Catherine T. MacArthur Foundation, Andrew W. Mellon Foundation, Edward John Noble Foundation, Pew Charitable Trusts, Lynn R. and Karl E. Prickett Fund, Rockefeller Brothers Fund, Surdna Foundation, Turner Foundation, U.N. Population Fund, Wallace Genetic Foundation, and Frank Weeden Foundation.

PUBLICATIONS of the Institute include the annual *State of the World*, which is now published in 27 languages; *Vital Signs*, an annual compendium of the global trends—environmental, economic, and social—that are shaping our future; the *Environmental Alert* book series; *World Watch* magazine; and the *Worldwatch Papers*. For more information on Worldwatch publications, write: Worldwatch Institute, 1776 Massachusetts Ave., N.W., Washington, DC 20036; or fax (202) 296-7365.

THE WORLDWATCH PAPERS provide in-depth, quantitative and qualitative analysis of the major issues affecting prospects for a sustainable society. The Papers are authored by members of the Worldwatch Institute research staff and reviewed by experts in the field. Published in five languages, they have been used as a concise and authoritative reference by governments, nongovernmental organizations, and educational institutions worldwide. For a partial list of available Papers, see page 68.

DATA from all graphs and tables contained in this book, as well as from those in all other Worldwatch publications of the past year, are available on diskette for use with Macintosh or IBM-compatible computers. This includes data from the *State of the World* series, *Vital Signs* series, Worldwatch Papers, *World Watch* magazine, and the *Environmental Alert* series. The data are formatted for use with spreadsheet software compatible with Lotus 1-2-3, including Quattro Pro, Excel, SuperCalc, and many others. Both 3 1/2" and 5 1/4" diskettes are supplied. To order, send check or money order for $89, or credit card number and expiration date (Visa and MasterCard only), to Worldwatch Institute, 1776 Massachusetts Ave., NW, Washington, DC 20036. Tel: 202-452-1999; Fax: 202-296-7365; Internet: wwpub@igc.apc.org.

Printed on 100-percent non-chlorine bleached, partially recycled paper.

Table of Contents

Sections of this paper may be reproduced in magazines and newspapers with written permission from the Worldwatch Institute. For information, call the Director of Communications, at (202) 452-1999 or Fax: (202) 296-7365.

The views expressed are those of the authors, and do not necessarily represent those of the Worldwatch Institute, its directors, officers, or staff, or of its funding organizations.

ACKNOWLEDGMENTS: Many people and organizations made this study possible. The Energy Foundation provided specific support for this work. The authors would also like to thank Paul Bartlett, Varis Bokalders, Mark Broyles, Nigel Howard, Frank Kensill, Lee Eng Lock, Peter Schmid, Lisa Surprenant, Brenda Vale, Robert Vale, and Alex Wilson for their comments on early drafts of this paper.

DAVID MALIN ROODMAN is a research associate at the Worldwatch Institute, where he studies the relationship between economic forces and environmental problems. He is coauthor of the Institute's annuals, *Vital Signs 1994* and *State of the World 1995*. He is a graduate of Harvard College, where he studied pure mathematics.

NICHOLAS LENSSEN is a senior researcher at the Worldwatch Institute, where he studies energy policy, alternative energy sources, nuclear power, radioactive waste, and global climate change. He is author or coauthor of four Worldwatch Papers, five of the Institute's annual *State of the World* reports, and *Power Surge: Guide to the Coming Energy Revolution* (W.W. Norton, 1994). He is also associate project director for *Vital Signs*.

Introduction

Modern buildings, like other artifacts of industrial civilization, represent an extraordinary achievement with a hidden cost. They make life easier for many today, but their construction and operation inflict grievous harm upon the environment, threatening to degrade the future habitability of the planet. Buildings account for one-sixth of the world's fresh water withdrawals, one-quarter of its wood harvest, and two-fifths of its material and energy flows. This massive resource use has massive side-effects: deforestation, air and water pollution, stratospheric ozone depletion, the risk of global warming. Moreover, up to 30 percent of new and renovated buildings suffer from "sick building syndrome," subjecting occupants—who spend up to 90 percent of their time indoors—to unhealthy air.[1]

Unlike pollution from cars and factories, which has been the subject of public battles in many parts of the world, the harm caused by buildings has largely escaped scrutiny. This is ironic, for the problems with buildings are easier to fix. There are cost-effective ways to avoid almost all of the damage that a new structure does, and still to preserve the security, comfort, and amenities that people expect of modern buildings. For example, air conditioners and refrigerators free of ozone-depleting chemicals are available in industrial countries. More impressive but less common are buildings that use just 2.5 percent of the heating energy that conventional ones do. Likewise, there are modern homes made of unbaked earthen blocks whose production gives off 0.2 percent of the pollution that brick making does.

What is encouraging, and perhaps surprising, is that buildings that are better for the environment are better for people.

Some examples show how:

- An affordable housing development in Dallas, Texas, has slashed utility bills by $450 a year per dwelling by incorporating solar heating and efficient appliances that add only $13 a year to mortgage payments.

- A new bank headquarters in Amsterdam uses 90 percent less energy per square meter than its predecessor. It cost $0.7 million extra to build but is saving $2.4 million a year. Employee absenteeism has also dropped, saving $1 million annually.

- A U.S. Postal Service facility in Reno, Nevada, spent $300,000 to improve lighting, and is saving $50,000 a year in electricity. Worker productivity has risen, saving $500,000 a year.

- Home values in a Davis, California, subdivision that incorporates solar heating and bike paths have risen 12 percent above those of conventional houses nearby.

- In a Dutch housing project emphasizing energy and water efficiency as well as the use of non-toxic materials, one resident, thankful that her child no longer suffers serious asthma attacks, says, "We want to stay here forever."[2]

Clearly, the building industry—designers, financiers, developers, and contractors—has a key role to play in the creation of a sustainable society. The important question is whether it will act fast enough. In the last hundred years, the amount of heat-trapping carbon dioxide in the air has risen 27 percent, of which one-quarter has come from the combustion of fossil fuels to provide energy for buildings. This build-up threatens ecosystems, agriculture, and human settlements with higher temperatures and changed weather patterns. Meanwhile, 20 percent of the earth's forests have disappeared. Around the world, mining of copper, bauxite, and iron ore resources for building materials continues, pouring large quantities of pollutants into nearby air and water. All of these trends are accelerating, and the damage they have done and may do is often irreversible.[3]

What makes the need for change in the building sector particularly urgent is that buildings last a long time. Once a structure is completed, it is harder and less economical to reduce its energy and water use and improve its air quality than it is to design from scratch for efficiency and health. In the face of uncertainty about problems like global warming, the precautionary principle—the simple, conservative idea that some risks are not worth taking—argues for action today rather than reaction tomorrow.

The construction and population booms in developing countries as diverse as Argentina, China, India, and Turkey further underscore the need to improve buildings. Roughly 2 billion people now live and work in resource-intensive buildings; in 50 years the number may reach 8 billion. If society does not change the way it makes modern homes and workplaces, already severe environmental problems may grow much worse. No structure, no matter how well built, could fully protect its residents from the effects.[4]

Around the world, the sprouting up of skyscrapers, highrise apartment blocks, and suburban-style homes presents a fleeting opportunity to reduce the toll buildings exact. Industrial countries need to ensure that as they make new buildings and renovate old ones, they avoid old mistakes. Developing countries can avoid copying nations that industrialized earlier and even leapfrog them by employing environmentally sound and healthful techniques and technologies, some of which are derived from their own indigenous architecture. Along the way, notions of what is progressive and what is primitive may be rearranged.

Since buildings are the business of everyone, all concerned need to contribute to minimizing the industry's impact on the environment and on human health. Governments, educators, investors, and consumers can help the building industry alter its course by formulating better policies and making better investment decisions. A concerted effort by all of these parties will ensure that people can provide shelter for themselves without jeopardizing the livability of their greater home—the planet.

Modern Buildings, Modern Problems

By all appearances, modern buildings are spectacular triumphs of the industrial age. Luxurious suburban homes from Stockholm to San Francisco and gleaming skyscrapers from Brasilia to Bangkok deliver myriad services that the planet's richest inhabitants view as necessities and the rest eagerly seek: indoor plumbing (including hot water on demand), precise climate control, lighting at the flip of a switch, refrigeration, communication, and even entertainment.

Once confined to a handful of industrial countries such as Germany, Japan, and the United States, the resource-intensive building is becoming a global phenomenon as cities grow and middle classes develop. Turkey, for example, witnessed a 13-fold rise in permitted building construction between 1963 and 1993. (See Figure 1.) New construction in South Korea jumped by a factor of nearly 50 over the same period. In countries where there used to be mostly small houses, barns, churches, and temples, new types of buildings are appearing: large stores, residential and office high-rises, even skyscrapers. Where wood, bamboo, brick, or unbaked earth once was the dominant building material, now concrete and steel are gaining ground. Where formerly buildings got their heat from the sun or from burning biomass, or captured the wind to keep cool, today structures with heaters and air conditioners are taking their place.[5]

Not only are there more modern buildings—some types are also getting bigger. For example, average home size has risen in industrial countries since World War II even as family size has shrunk, a trend that the oil shocks of the seventies failed to slow. (See Figure 2.) In the United States, floor space per person more than doubled in new single-family houses between 1949

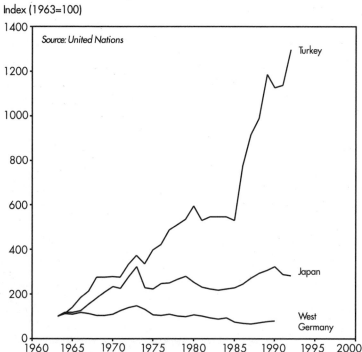

FIGURE 1

Building Construction Activity in Turkey and Japan, 1963–92, and in West Germany, 1963–90

Index (1963=100)

Source: United Nations

and 1993. As Gopal Ahluwalia of the National Association of Home Builders in the United States explains (only half facetiously), "everybody wants a media room, a home office, an exercise room, three bathrooms, a family room, a living room, and a huge, beautiful, eat-in kitchen that nobody cooks in."[6]

Modern buildings have given consumers more of what they want in terms of floor space and special features. Yet in less visible ways, buildings are less desirable than they could be. They use energy and water inefficiently, and they often wear out quickly, wasting natural resources, money, and human labor. Many also create unhealthy indoor air, making people feel unwell at times and possibly contributing to cancer and immune disorders.

Residential Floor Space Per Person, Selected Countries, 1973-90

The roots of the problems caused by building lie in the industrial revolution, which made invisible much that used to be visible. Mechanization, which supplants human labor with energy use, and specialization, which exploits the human ability to become more efficient at tasks through practice, have each made individual workers vastly more productive. Together, they have given birth to a diverse economy that can provide customers from New Delhi to Los Angeles with everything from tropical hardwoods to rooftop air conditioners. In this global system, the impacts of any one person's actions on the planet can be widespread, but confinement to narrowing economic niches has made it harder for building industry workers to understand the world beyond their jobs.[7]

The building industry has used machines and job specialization to scale up its work, cut costs, and produce the remarkable buildings of today, but at great environmental cost.
Bulldozers and drills have replaced saws and pickaxes, allowing
loggers and miners to work at massively destructive rates, while
ships and trains now transport lumber and metal, coal and oil
by the ton to construction sites and power plants far away.
Chemists and metallurgists have found ways to add to the cor-
nucopia by turning naturally occurring substances into pollution-
intensive, mass-produced materials such as steel, cement, and
plastics. As builders have gained access to new materials, distance
has insulated them from the environmental consequences of
their choices, facilitating destructive decisions.

Once the job of a few generalists, the process of making and
maintaining a building now encompasses a score of roles, including miner, logger, shipper, supplier, developer, financier, architect, engineer, general contractor, subcontractor, inspector, and
building manager. As the complexity of construction has risen
and individual actors have become distanced from each other
and from either end of the process—the environment at one end,
and the people who receive the finished product at the other—
they have naturally focussed on their immediate, day-to-day
concerns, whether they be mining a resource as quickly as possible, minimizing up-front costs, maximizing a commission, or
meeting a deadline.[8]

This separation has made it hard for designers to think about
buildings as a whole, leading to profligate resource use. For
example, most designers set key building parameters—such as
shape, location of windows, and amount of lighting—without
concern for how their decisions can substantially affect energy
use down the road. When the engineers receive the blueprints,
they have to specify a large enough climate control system to
compensate for this lack of foresight, and the opportunity for
savings is lost. "[It is] as if the effort were not a team play but
a relay race," writes Amory Lovins, research director of the Rocky
Mountain Institute in Snowmass, Colorado.[9]

The greatest fissure to appear in the building process has
divided the occupants from the building industry, and has result-

ed in buildings that wear out sooner than they should and waste large amounts of resources. Most people and organizations do not build their own facilities but buy them on the open market. Yet few products are as expensive and complex as a building. This creates a complicated tension between buyer and producer.

On the one hand, an industry naturally develops its own conservative culture when one unsold building or unpopular development can spell financial disaster, and one unsound structure, human catastrophe. By doing things the way they always have and the way everybody else always has, industry actors reduce the risk of failure, but at the expense of genuine variety and innovation. (In the litigious United States, fear of liability lawsuits also encourages designers and builders to stick with standard practice.) A sort of informal cultural monopoly results.

Buyers, on the other hand, come to the market with a host of concerns: not just price and location, but long-term costs, resalability, durability, appearance, and amenities—from jacuzzis to computer network wiring. Because the buildings available to them are so much alike, buyers can rarely find exactly what they want, so they focus on what is most important and easiest to measure: usually price, location, size, and a few standard features. Customers may get what they want in these respects, but they inadvertently end up giving the industry leeway to cut corners on energy efficiency, indoor air quality, and durability.

In centrally planned economies, the situation is worse. People lose nearly all power over what they inhabit, resulting in even more wasteful buildings. Lacking thermostats, Russians often resort to opening windows to cool down overheated apartments. Similarly, in the mid-1980s, Chinese buildings used three times as much energy for heating as comparable U.S. buildings, even though inside temperatures remained lower.[10]

It is harder to document systematically that buildings are often poorly made, but the extreme cases of natural disasters cast the problem in sharp relief. In one day in August 1992, for example, Hurricane Andrew damaged or destroyed more than 100,000 homes on the southern tip of Florida, disrupting the lives of thousands of families and inflicting approximately $30 billion in property damage. Amid the wreckage, however, some homes stood largely intact. Post mortems revealed that widespread and undetected building code violations were responsible for the loss of many homes. Thus much of the housing in the fastest-growing U.S. state seems to have been sloppily built. And, because of the rebuilding activity that ensued, Andrew cut other swaths through the American Southeast and Northwest, where thousands of hectares were logged to provide lumber.[11]

As devastating as Hurricane Andrew was, the earthquake that rocked Spitak, Armenia, in 1988 was far deadlier. In a matter of seconds, the temblor wrenched apart schools and apartment buildings, killing more than 25,000 people. Post-disaster analysis revealed a pattern governing which structures survived. Soviet workers had constructed most of the buildings of nine stories or less as sturdy concrete boxes, binding the load-bearing walls tightly to each other and to the floors. These came through the earthquake largely unscathed. The taller ones, however, were built as loosely jointed beam-and-column frames with non-load-bearing walls, and they collapsed like houses of cards.[12]

Lacking thermostats, Russians often resort to opening windows to cool down overheated apartments.

Yet for the building industry to understand fully what it is up against in making buildings last, it needs to appreciate not only how the world has changed, but that it continues changing—and rapidly. All too frequently, buildings that do not adapt to change are abandoned or demolished.

Rapid movement and shifting needs strain the building stock. As decades pass, whole industries rise, fall, and move quickly across the landscape. And people follow. In the United States—a huge free trade and travel zone—one in six people moves each year, often thousands of kilometers. As international trade and migration limits gradually disappear, for example within the European Union, people elsewhere will likely relocate more too. Unfortunately, occupants who expect to move soon tend to invest less in structures for the long term. Only one in three American homes is well maintained, and that figure may be falling, according to some estimates.[13]

Dramatic context changes—such as land values doubling or whole neighborhoods jumping to the suburbs—can take an even greater toll. Sixty percent of West Germany's buildings survived the onslaught of World War II, yet less than 15 percent of those left standing weathered the rapid development of the ensuing 30 years. Skyrocketing land values in postwar Tokyo translated into a churning real estate market in which, by the eighties, new buildings lasted only 17 years on average before being replaced with taller ones.[14]

When rapid transitions occur between traditional and contemporary buildings, they also dramatize the differences between the two types. In Beijing, where the government is razing many old neighborhoods of tightly packed one-story dwellings in favor of new shopping centers and high-rises, police have sometimes had to precede bulldozers in order to evict occupants who are increasingly resistant to moving. Urs Morf, a Swiss journalist, describes the shifting attitudes towards the new buildings among residents he talked to since resettlement began in the early 1980s:

> Back then...the prospect of running water, central heating, more living space, and above all, more personal freedom, lured prospective tenants away from the constriction and constant mutual surveillance of traditional, multi-generational living quarters. Since then, however, word has gotten round of the many disadvantages of life in the new housing blocks, which were built as cheaply as possible and deteriorate very rapidly. These days the additional freedom and space are perceived as anonymity, which, as everywhere in the world, has brought with it a tremendous increase in petty criminality and eliminated the old sense of neighborly solidarity. Home and work space are now separated by great distances, and the once-compact family group has been atomized.[15]

The Beijing experience shows that buildings fall short if they only provide comfort, security, and modern amenities. To the extent that modern structures have become destructive to the world around them, they can be seen as more primitive than traditional structures. Although they have improved in many ways since the industrial revolution, buildings will have to evolve much further if they are to answer all human needs.

Quality Construction

As the story about Beijing and the disasters in Florida and Armenia suggest, *quality* is one of the keys to ecological design. Environmentally sustainable buildings need to excel in many ways, from indoor air quality and energy efficiency to durability and flexibility. Mechanization and specialization, though, have largely replaced craft with a zeal for speed and an assembly-line mentality. Thus one of the great questions of sustainable design today is how best to exploit the power of machines and the efficiency of job specialization without sacrificing high standards.

In some ways, the building industry today finds itself in the same position that another conservative, capital-intensive industry did in the 1970s. At the time, west European and North American automobile makers produced cars of moderate quality at moderate prices. Though buyers in these countries would have preferred better ones at lower cost, they were powerless to extract them from the manufacturers—that is, until Japanese companies provided an alternative. Consumers then started voting with their buying choices, giving domestic producers no option but to begin learning from the Japanese. Likewise today, builders could gain an edge by marketing higher-quality products that emphasize environmental and health concerns.

Another country that many builders could learn from is Sweden, which has earned an international reputation for the quality of its houses. Juxtaposed with the United States, the contrast is striking. In the United States, 95 percent of new houses include some factory-assembled components. High-volume assembly lines, staffed by low-wage, unskilled workers, typically produce these parts, which rarely meet the exacting dimensional standards needed to make homes airtight and highly energy efficient. In Sweden, where most homes are also made of wood, industry has found a way around this dilemma. House component factories there are smaller, and are run by groups of highly skilled workers who use semiautomatic tilt-tables to hold wall panels in place while they add framing, windows, insulation, and vapor barriers. With custom tools and machines doing the muscle work, people can concentrate on doing what they do best and bring craft back into their work—painstakingly making parts fit well—without raising costs. The Swedish "factory crafting" system produces such high-efficiency, high-quality homes that the country actually *exports* houses.[16]

As impressive as it is, the Swedish innovation is only one part of the solution to a much larger puzzle. Since so many of the problems with buildings arise from disconnections between the environment, customers, and various participants in the construction process, it seems logical that the entire series of operations must be re-engineered to facilitate communication, to assist people in seeing the connections between what they do and the world around them. Teams that include all the actors can reknit the disparate parts of the building process, helping them take into account the effects of their work on each other and the environment. "Integrative design," as this is known, does not imply rejection of industrialism. Rather, it is a new way to harness it, a way to employ machines and specialized knowledge in the service of quality. Designers who have delved into the challenges of ecological construction have discovered that integrative design is critical to success. "Before you design a building, you have to redesign the building process," says John Picard, a Los Angeles-based consultant in environmental design.[17]

For example, in 1978, the directors of what it now known as the Internationale Nederlanden (ING) Bank decided to build a new headquarters in Amsterdam. They chose to make it an "organic" building that was efficient in its use of energy and other natural resources, that was healthy for its users, and that integrated natural shapes, green plants, and art into a space that celebrated the human spirit. Early in the design process, the bank brought engineers, an architect, interior and landscape designers, scientists, and future occupants together into a working group.[18]

Instead of falling into the common pattern in which one person's solution becomes another's problem, the team worked to find solutions that satisfied several needs and wants simultaneously. For instance, they found a design that satisfied the workers' desire for operable windows while preventing large heat losses in winter or gains in summer. Once it went into operation in 1987, the new headquarters used a fifth of the energy of a new office building nearby for the amount of floor space, and less than a tenth as much as the previous headquarters. Among the other measures that permitted such reductions were good insulation and optimal use of natural lighting to supplant electric light (no desk is more than six meters from a window, for example). Energy bill savings covered the additional costs of the design in just four months of operation, and the new building proved to be such a healthy and pleasant place to work that absenteeism dropped 15 percent.[19]

It seems logical that the entire building process must be re-engineered to facilitate communication.

The need to make buildings flexible adds another group of concerns for the integrative design team. Although it is hard to completely immunize any building against upheaval, many can be made adaptable to more gradual changes. U.K. architect Francis Duffy, an expert on how structures evolve over time, has found that among buildings that do survive, exteriors tend to change every 20 years or so, while new wiring, plumbing, and climate control systems might be added every 7–15 years, and floor plans can change as often as every 3 years. From this perspective, buildings consist of several layers, each evolving at different rates.[20]

The key to making buildings last and adapt is to separate these layers clearly so that slow-changing ones do not impede alterations to fast-changing ones, according to writer and strategic planning consultant Stewart Brand. Developers of high-turnover commercial buildings, for example, build them with a load-bearing frame and internal walls that can be knocked down as required, allowing new occupants to tailor the floor plan to their own needs.[21]

Two neighboring buildings on the Massachusetts Institute of Technology campus in Cambridge, Massachusetts, illustrate the advantages of flexible buildings. One, the famous "Media Lab," was designed by the internationally renowned architect I.M. Pei, who by his own admission prefers to make buildings whose uses will not change over time. Although the 1985 structure is visually striking as sculpture, its cast-in-concrete fixity ill suits it for research in rapidly changing technologies. Indeed, some special-purpose rooms became obsolete by the time researchers moved in, and are now useful for little else. Upgrading the building's infrastructure has proved expensive because it entails taking a jackhammer to its concrete internal walls.[22]

Across the street stands a structure that could not be more different. "Building 20" was thrown together quickly during World War II to house a crash program in radar development. The university has slated it for demolition ever since, but its sheer adaptability—for example, the ease with which one can punch a hole for cables through one of its wooden walls—has made it an excellent incubator for innovative ideas, and repeatedly saved it from the wrecking ball. (The companies that invented the minicomputer and the predecessor of the Internet both started there.)[23]

One recent New York City project demonstrates the dual advantages of integrative design and adaptable buildings. After the National Audubon Society decided to move its headquarters, its interdisciplinary design team settled on refurbishing a century-old, eight-story brownstone in lower Manhattan, rather than start a new building from scratch in the suburbs. By rescuing the old one from demolition, Audubon calculates that it kept some 200 tons of steel, 9,000 tons of masonry, and 560 tons of concrete out of the dump. The team then drew on new technologies—including an efficient cooling system, insulating windows, and a good ventilation system—to cut energy use by 60 percent, reduce water use, and improve air quality. Completed in 1992, the building's improvements will pay for themselves within five years, and will continue generating savings thereafter. Features like daylighting and much fresher air have also made the office a more pleasant and productive place to work.[24]

Although integrative building design may seem novel in the West, it is really an old idea in a new guise. It was in borrowing the management ideas of the American W. Edwards Deming that many Japanese manufacturers of cars and other products began proving decades ago that a team-oriented management structure is the best way to ensure quality in industrial production. In fact, major Japanese construction firms already combine design, engineering, construction, maintenance, and building operation services under one roof, although, as in most countries, only a few have begun to think about the broad environmental and health impacts of their work.[25]

One exception is the $18 billion Shimizu Corporation, the largest construction company in Japan, and one of the five largest in the world. In 1991, it adopted a Global Environmental Charter, in which it pledged to incorporate environmental concerns into all of its work. Although rhetoric usually precedes reality, the company has made significant improvements. For example, it has developed an advanced robotic system for constructing high-rises that allows for the just-in-time delivery of pre-cut building materials to the site each day. Workers need not store components before they use them and do not have to cut pieces to size on site, thus reducing packaging and construction waste by up to 70 percent.[26]

In Germany, almost every major city now has a store that sells a wide variety of healthy building materials.

In the mid-1990s, there are signs that interest in ecological building has begun to increase in many countries. In Germany, almost every major city now has a store that sells a wide variety of healthy building materials. And the largest architectural and engineering firm in the United States—Hellmuth, Obata, and Kassabaum—has begun systematically evaluating the environmental and health impacts of the materials that it specifies. In the United Kingdom, a government-industry collaboration resulted in the 1994 publication of an environmental code of practice for building professionals by the Building Services Research and Information Association (BSRIA). It sold 700 copies—more than any of BSRIA's other publications—in the first three months alone.[27]

The growing concern within the building industry about human and environmental health gives reason to hope that more companies will eventually use integrative design to make better buildings. If they do, these market leaders will be putting industrialism to use in solving problems that it helped create. Like the Japanese auto makers before them, they will advance their own interests and those of consumers as well.

More Than Skin Deep

How buildings look reflects how they are made. Traditional architectures derived much of their form from the unforgiving limits of local materials, the vagaries of local climate, and the need to use the one to buffer against the other. From the spreading red rooftops of Florence to the striking black tents of the Bedouin nomads, indigenous shelter often approached organic beauty. As Frank Lloyd Wright observed:

> The true basis for the most serious study of the art of architecture lies with those indigenous more humble buildings everywhere, that are to architecture what folklore is to literature, or folksong to music, and with which academic architects are seldom concerned...Functions are truthfully conceived and rendered invariably with natural feeling. Results are often beautiful and always instructive.[28]

As contemporary construction has become divorced from ecological principles, its aesthetic character has changed radically. A confluence of forces and limits still shapes buildings, but it originates in offices of businesses and governments far away. As a result, modern buildings rarely respond to their immediate context the way traditional ones do, nor do they offer visitors a sense of connection to the fabric of a place.

In the case of residential buildings, developers have gone from constructing homes one at a time to entire neighborhoods in which any variation is superficial. Today there are ranch-style houses in Provence and Oregon that look remarkably similar, and virtually identical residential high-rises from Philadelphia to Budapest to Singapore. In the case of larger office and apartment buildings made from concrete and steel, economics has driven a trend away from ornamentation and toward simple geometric forms, usually boxes, that are quick to design and build. The major exceptions to this uniformity have appeared where there is profit in standing out—in stores and restaurants designed to catch the eye of motorists speeding by. Taken as a whole, the settled landscape worldwide is coming to consist of little more than monotonous business and residential neighborhoods scored with garish commercial strips.[29]

As designers begin to solve the environmental and health problems generated by buildings, allowing the immediate physical and human context to impose form to a much greater degree, buildings may once again become more beautiful. Several human and environmental concerns shaped the new ING Bank headquarters, for instance. To help workers remain oriented as they moved around the building, designers determined that no part of it should look like any other. They also wanted to maximize exterior area to bring in daylight and to include courtyards with gardens visible from indoors. Together these priorities gave rise to an irregularly shaped building that snakes around its lot in a jagged S, standing out sharply from its rectilinear neighbors. Viewed from above, it looks like a living organism surrounded by inert boxes.[30]

As Wright suggests, contemporary designers have something to learn from ancient builders, but they need not copy them slavishly, nor sacrifice modern amenities. Rather, they can extract a set of guiding *principles* of ecological design and apply them in the modern context. As they do this, modern buildings too may begin to develop their own recognizable aesthetic language, flowering from region to region into dozens of modern vernaculars.[31]

Construction Destruction

A modern environmental architecture will have to respond to the modern context, which is far more complicated than the one within which traditional builders worked. Design decisions today contribute not only to local environmental problems, but to regional and global ones, and to health problems as well. (See Table 1.)

Buildings first make their presence felt during construction. Construction workers turn some 3 billion tons of raw materials—40 percent of the total flow into the global economy—into buildings each year. Almost all of this is essentially dirt—clay for bricks, and gravel and sand for concrete. Quarrying these materials usually has few off-site impacts, but on site it can obliterate foliage and scar the earth.[32]

Though the tonnages involved are smaller, using metals and plastics has far more environmental impact than using quarried materials because it entails either purification from low-grade ores or heavy chemical processing. For example, of the copper employed in U.S. buildings (nearly half of the total used in the country), some is recycled material, but 80 percent is extracted from irreplaceable virgin ores and purified through a process that is one of the largest sources of air pollution in the country. Polyvinyl chloride, better known as vinyl or PVC—a chlorinated plastic that is widely used in piping, siding, and windows—is little better. It is difficult to recycle, and its production and incineration (if that is how it is disposed of) generate carcinogenic dioxins, vinyl chloride monomers, and other pollutants. Germany's Health Ministry and the American Public Health Association, among other institutions, have called for phasing out PVC where viable substitutes exist.[33]

In the industrial countries of North America, Scandinavia, and the Pacific, wood remains the material of choice for houses. As a result, construction accounts for more than a quarter of the world's annual 3.5-billion-cubic-meter appetite for wood. (Additionally, 55 percent of the world's wood harvest is burned to cook food and heat homes, primarily in developing countries.)[34]

TABLE 1

Impacts of Modern Buildings on People and the Environment

Problem	Buildings' Share of Problem	Effects
Use of Virgin Minerals	40 percent of raw stone, gravel, and sand; comparable share of other processed materials such as steel	Landscape destruction, toxic runoff from mines and tailings, deforestation, air and water pollution from processing
Use of Virgin Wood	25 percent for construction	Deforestation, flooding, siltation, biological and cultural diversity losses
Use of Energy Resources	40 percent of total energy use	Local air pollution, acid rain, damming of rivers, nuclear waste, risk of global warming
Use of Water	16 percent of total water withdrawals	Water pollution; competes with agriculture and ecosystems for water
Production of Waste	Comparable in industrial countries to municipal solid waste generation	Landfill problems, such as leaching of heavy metals and water pollution
Unhealthy Indoor Air	Poor air quality in 30 percent of new and renovated buildings	Higher incidence of sickness—lost productivity in tens of billions annually

Source: Worldwatch Institute, based on sources cited in text.

In the last hundred years, global forest cover has shrunk by a fifth; more than half of what remains consists of isolated forest fragments or commercial monoculture stands. These changes have eliminated thousands of plant and animal species and have destroyed the homelands of many indigenous peoples. Most timber harvesting in old-growth forests is really timber mining: it is far above sustainable levels, and must eventually fall off. Symptomatic of the spreading forestry crisis, many nations have exceeded their domestic capacity to supply wood and have taken to importing large amounts. Japan, by far the world's largest importer, fed its building boom through the eighties partly with timber from dwindling Pacific Rim forests.[35]

Putting structures up also produces large amounts of solid waste, as does tearing them down. Erecting a typical, 160-square-meter, 150-ton home in the United States generates some 7 tons of refuse. And for every six houses or apartment units constructed, one falls to the wrecking ball—about 150,000 each year. As a result, the country generates roughly as much construction and demolition waste as municipal garbage; the European Union produces 50 percent more.[36]

Making buildings also consumes energy, from mine to foundry to construction site. Steel, glass, and brick require large amounts of fossil fuels for high-temperature production. Transporting materials to the building site takes still more energy. In the United States, these activities account for roughly 9 percent of energy use.[37]

Buildings in operation do far more damage than buildings under construction, drawing heavily on energy and clean water flows. Between 1971 and 1992, primary energy use in buildings worldwide grew an average 2 percent annually. In 1992, their share of total energy use stood at 34 percent. This included 25 percent of fossil fuels, 44 percent of hydropower, and 50 percent of nuclear power.[38]

Adding in the fuels and power used in construction, buildings consume at least 40 percent of the world's energy. They thus account for about a third of the emissions of heat-trapping carbon dioxide from fossil fuel burning, and two-fifths of acid-rain-causing sulfur dioxide and nitrogen oxides. Buildings also contribute to other side effects of energy use—oil spills, nuclear waste generation, river damming, toxic run-off from coal mines, and mercury emissions from coal burning.[39]

A survey of water use in buildings tells a similar story. From Beijing to Los Angeles, growth in urban water use is lowering water tables and necessitating large projects that siphon supplies away from agriculture. In addition, electric power plants use water as a coolant, much of which drains into rivers, carrying thermal and chemical pollution. These two uses contribute about equally to buildings' one-sixth share of global water withdrawals.[40]

Finally, many modern buildings also create dangerous indoor environments for their inhabitants. Indeed, the estimate that "sick building syndrome" occurs in 30 percent of new or renovated buildings worldwide may be conservative. Ventilation systems installed to protect air quality often hurt it, subjecting occupants to stale air for hours on end, or harboring and spreading unhealthy molds. Headaches and nausea can result. Sealed buildings also trap volatile organic compounds (VOCs) that can seep out from composite materials, furniture, carpets, and paint, and accumulate at concentrations hundreds of times higher than those just outside. Long-term exposure to some VOCs may increase the risk of cancer and immune disorders.[41]

Ventilation systems installed to protect air quality often hurt it.

The medical and worker productivity costs of unhealthy indoor air may run into the tens of billions of dollars each year. In addition, some researchers suspect that forced air circulation may facilitate the spread of airborne illnesses like the common cold and influenza. If these suspicions are correct, the economic impact of sick buildings could run to hundreds of billions of dollars annually.[42]

How can buildings be made better? Some answers will come from the past—in an era of apparent plenty, it is easy to forget that scarcity has often been the mother of invention, spurring myriad cultures over thousands of years to perfect ways to do more with less. Other answers will emerge from laboratories, where researchers continue to develop healthier, more resource-efficient materials and technologies. Increasingly, successful building will recover some of the ancient techniques recently discarded by designers and combine them with new technologies to create a synthesis that is better for the environment, and better for humanity.

Material Concerns

One basic decision a builder must make is what to build with. The choices can be bewildering, ranging from traditional adobe (sun-dried mud bricks) to modern concoctions like vinyl. In ordering from this menu, each designer judges materials according to a large set of criteria: in addition to being easy to work with, they should insulate; block air leaks; provide strength in compression and tension; resist fire, moisture damage, and biodegradation; look good; and not cost too much.

The ecologically concerned designer has even more to worry about. Behind each material lies a manufacturing history, often quite long. Each step taken to move or make the material carries an environmental or health cost, whether exacted by an ax or a tailpipe. For materials that emit volatile organic compounds, that history continues right inside the building. For almost every material used in industrial countries, the price is high, and apparent alternatives are few.

The building industry can use materials much more sustainably than it does. The guidelines for doing this are well known: Look for substances that entail a minimum of transportation and processing—in other words, local, natural materials. Avoid as much as possible materials that emit toxins. Use materials efficiently. Use ones that are renewable, recyclable, or both.

To apply the first rule, it is helpful for designers to know how much energy it takes to make a material. The less processing it undergoes, and the shorter the distance it is transported, the less energy it uses, and the less pollution it creates. Because relatively little energy is needed to obtain traditional materials such as wood, stone, and adobe, they all cluster at the low-impact end of the materials spectrum. The substances whose strength, transparency, and electrical conductance make modern buildings possible, however, are more energy- and pollution-intensive. (See Table 2.)

Steel production, for example, can be highly polluting. Iron mining produces tailings that can leach heavy metals into nearby streams; and open-hearth steel making can emit lead and

TABLE 2

Energy Used in Production and Recycling of Selected Building Materials, United Kingdom[1]

Material	Virgin Production	Recycling
	(gigajoules per ton)	
Concrete	0.5–1.5	0.5–1.5[2]
Brick[3]	2.5–6.1	
Wood (Domestically Harvested)[3]	4–5	
Glass	13–25	10–20
Plastics	80–220	50–160
Steel	25–45	9–15
Copper	70–170	10–80
Aluminum	150–220	10–15

[1]The wide ranges in these data reflect variations in manufacturing processes from plant to plant, and in distances materials are transported. [2]Using old, crushed concrete for aggregate and all new Portland cement. [3]Generally not recycled, though can be reused or made into other building products.

Sources: Nigel Howard, Davis Langdon Consultancy, London, printout, September 20, 1994, and private communication, September 27, 1994; recycling energy use for glass and plastics are Worldwatch estimates, based on ibid., and on data from Jeffrey Morris and Diana Canzoneri, *Recycling Versus Incineration: An Energy Conservation Analysis* (Seattle, Wash.: Sound Resource Management Group, 1992).

other poisonous heavy metals. Open-hearth steelworks are common in the former Eastern bloc and some developing countries, but fortunately, in developed market economies, more advanced furnaces are the norm, using more recycled steel, consuming less energy, and generating less pollution. Still, according to U.K. data, the use of materials such as steel, copper, aluminum, and concrete makes each square meter of floor space in a large office building 2–4 times as energy-intensive—and therefore probably 2–4 times as pollution-intensive—as a house. Mindful of their special properties and high environmental costs, the environmentally concerned designer needs to use modern materials the way a jeweler uses precious metals—sparingly.[43]

Returning to the idea that buildings should be both durable and flexible in design leads to another important distinction among materials. Inorganic masonry tends to be inflexible but durable, making it an ideal structural material. Almost all surviving ancient buildings—the Mayan and Egyptian pyramids, the Parthenon, the Taj Mahal, for example—were made of stone. In contrast, organic materials, notably wood, are more perishable. The complex molecules in a wooden beam contain 15 gigajoules of energy per ton, making it potential food for rot and termites, or fuel for fire. Nevertheless, since wood is easier and cheaper to work with, it may be better suited than masonry for building components that change every few years or so, such as non-load-bearing walls.[44]

Cultures that have developed in climates where organic materials are sparse have long made their dwellings out of a durable, low-impact, inorganic material: the earth itself. Ancient Chinese builders rammed mud into rigid molds to make the Great Wall, and archaeologists have found millennia-old adobes in the Indus Valley and the Middle East. Much of Europe, from England to the Balkans, turned to adobe and rammed earth after razing its own forests in the Middle Ages. Today, roughly two-fifths of humanity lives in earthen dwellings.[45]

The long history of earthen materials makes them seem primitive, but they are fully appropriate for today's small and medium-sized buildings. Unlike cement and brick, they do not require energy-intensive, high-temperature firing. Soil suitable for building lies underfoot at construction sites in large parts of the Americas, Europe, Africa, and Asia. In Yemen, earthen buildings reach five stories in height. Earthen structures are naturally fire- and rot-proof, and with proper technique can be as earthquake resistant as concrete and brick buildings.[46]

Builders in regions where the local soils are unsuited for building materials may on balance prefer perishable but sustainably harvested organic materials to inorganic ones shipped from far away. Although wood will probably remain popular in places where it is widely used today, other organic materials are available. Nebraska-style homes in the United States, made from stacked and plastered straw bales, have garnered renewed interest among some builders in the early 1990s. In addition to being plentiful (farmers burn 180 million tons of straw a year in the United States alone, enough for 5 million houses), straw is easy to work with, and it is an excellent insulator.[47]

In industrial countries, where labor is expensive, users of materials such as mud and straw have been mechanizing to bring down costs. For half a century, one U.K. company has been making large structural panels out of compressed straw, capable of replacing wood beams in roofs, ceilings, floors, and walls. The company has exported factories to some 50 countries, including Australia, the United States, Kenya, Brazil, and China. Each plant puts out 2,000 square meters of "stramit" paneling a day with a minimum of labor, energy, and pollution. The firm reports that interest has picked up noticeably in recent years. Similarly, companies in Australia, Germany, Switzerland, and the United States sell small machines that make up to 900 compressed earth blocks per hour. Michael Langley, a builder in Austin, Texas, reports that by using one of these machines he can compete effectively with makers of more conventional stone or brick homes. The ecological price of earthen blocks made this way—as measured by energy use—is one-five-hundredth that for bricks.[48]

Another concern with materials is their potential impact on indoor air quality. Most bonding and drying agents in carpets, veneers, particle board, plywood, and petroleum-based paints emit health-threatening volatile organic compounds. For paints, there are good linseed-oil-based alternatives. And for plywood, strand board, and particle board containing standard formaldehyde-based glues, there are a variety of alternative options, including boards made with low-VOC glues, cement, gypsum, or lignin, a naturally occurring bonding agent in wood. In the long run, builders can also consider switching to inorganic materials, which cause few air quality problems.[49]

Having chosen materials, the second challenge for builders is to use them as efficiently as possible. The art of making wood go farther is perhaps the most advanced because primary forests are already disappearing in many places. In putting up a house in Davis, California, in 1993, the Davis Energy Group and the

Pacific Gas and Electric Company (PG&E) found that they could space structural timbers further apart without threatening safety, cutting wood use in half. The Center for Resourceful Building Technology in Missoula, Montana, built a home a year earlier to demonstrate more sophisticated techniques for building efficiently with wood. Walls had insulation sandwiched between low-VOC strand board panels, instead of normal timber framing, which requires long beams from mature trees. In addition, the floor incorporated I beams made from wood scraps shaped like their steel counterparts to leave material only where it provides the most strength, and cutting wood needs by 75 percent. Such I beams have caught on with builders in the last 10 years. Like plywood, they are cheap because they are made of wood scraps that would otherwise be wasted. According to one study, 64 percent of U.S. home builders now use them.[50]

Builders can also use modern materials more efficiently. Steel use in new office buildings has fallen by two-thirds in the United States since the sixties, in favor of less-energy-intensive steel-reinforced concrete. In 1993, a French company developed a concrete that contains thousands of hair-sized steel wires instead of standard reinforcing bars. It is strong enough to do the work of conventional concrete that is three to four times as thick, and may eventually allow additional material savings.[51]

An equally important way to make materials go further is to reuse and recycle them. The first opportunity comes at the beginning of a building's life. Typical construction in North America generates 20–35 kilograms of solid waste per square meter of floor space, much of it leftover bricks, concrete, and wood scraps that are reusable or recyclable. Aware of these facts, the Canadian Energy Ministry launched a design competition in 1991 known as the Advanced Houses Program, to showcase houses that pushed the state of the art along environmental and health dimensions such as energy, water and materials use, and indoor air quality. Several of the 10 winners made impressive strides in cutting construction waste. Makers of the "EnviroHome" in Nova Scotia recycled two-thirds of the 3.4 tons of debris they would otherwise have sent to the landfill. And builders of the Waterloo, Ontario, "Green Home" generated an extraordinary five kilograms of waste—the rest of what ordinarily would have been discarded they either used in other buildings or recycled.[52]

Inorganic materials are usually easier to reuse than organic ones. Bricks and concrete blocks can easily be rescued intact from a building destined for demolition, and can last for millennia. St. Albans Abbey, still standing in southern England, was built 900 years ago partly out of bricks from nearby Roman ruins almost 1,000 years older. Modern-day demolition crews can crush large concrete blocks and mix the resulting powder with additional cement to make new concrete. A crew in Sydney, Australia, that recently knocked down an entire skyscraper, actually sorted the glass, steel, and concrete and sent it all off to the recyclers, a practice that could become common as landfilling becomes more costly.[53]

St. Albans Abbey was built 900 years ago with some bricks from nearby Roman ruins almost 1,000 years older.

It is harder for companies to find new customers for short, nail-filled wood timbers than for masonry blocks. However, in a process similar to paper recycling, they can make old wood products into new ones by grinding them down and reusing the fiber. One start-up company in California makes a material called "gridcore" from old newspapers, cardboard boxes, and timbers, molding it to maximize strength while minimizing weight. It believes that it will soon be possible to build entire homes with these new beams, studs, and panels. Already, the company has picked up some high-profile customers, including the electronics giant, Sony, and has begun collaborating with Armstrong World Industries, which makes 80 percent of the world's ceiling tiles.[54]

Materials choices are among the most complicated an environmentally concerned designer will face. The diversity of functional and environmental concerns surrounding materials is matched by the diversity of approaches to dealing with them. Given this complexity, there may be no perfect alternatives to conventional materials, only better ones. With choices evolving rapidly, designers should view materials selection itself as a continually changing process. Over time their understanding of the choices will improve, and new options like gridcore and earthen block machines will appear.

Designing with Climate

The Davis House—the California home built with half the wood of its neighbors—is remarkable for more than its way with materials. With its prominent two-car garage, and its neatly mowed front lawn, the home blends in seamlessly with its newly built suburban surroundings. But in a region where summer temperatures can top 40 degrees Celsius, it stands out in one subtle detail: it has no air conditioner.[55]

PG&E's main purpose in commissioning the Davis House was to demonstrate how good design can reduce energy needs and save money without sacrificing amenity. Using an integrative process, the designers decided they could save the most money and energy by avoiding add-on machinery as much as possible and instead adapting the house better to its immediate environment. Thus they gave it a minimally indented floor plan to reduce the amount of heat-absorbing surface area; they insulated it well; they positioned the windows to capture the sun's rays only at certain times of day and year; they specified a light-colored roof to bounce sunlight back into the sky; and they included tile floors and thick internal walls to soak up heat during the day and release it at night.[56]

The result was a design that strove to harness natural forces at times, while deflecting them at others, in order to create an indoor climate that was livable year round. Only then did the designers turn to machinery, adding both whole-house and room fans for the hottest days of the year and a super-efficient gas heater for the coldest.[57]

According to reviews from the house's residents, and objective measurements as well, PG&E has succeeded in creating a comfortable, efficient, and economical home. Preliminary data indicate that it is coming close to its goal of reducing energy use 60 percent, which means that it should save $1,600 on energy bills in its first 20 years. Once such techniques are commercialized, such a house should cost $1,800 less to build than a standard design would have.[58]

The kind of fit with local environment demonstrated by the Davis House is exceptional in industrial countries. Whereas designers traditionally sited buildings close to water supplies or designed them to capture prevailing breezes and the winter sun, most contemporary architects plan their structures according to other expedients—confident that hidden, energy-guzzling mechanical systems will compensate for what would otherwise be impractical design. Without these systems, buildings would be too cold in the winter and too hot in the summer.

Once mechanical climate control became practical, developers could put any type of building anyplace, so long as they greatly amplified the energy flow into it. For instance, the Cape Cod house, refined over generations to resist the powerful winds and exploit the weak winter sunlight found in the Northeastern United States, can now be plopped down in the milder climate of Virginia—and is, with a power-hungry air conditioner tacked on. Le Corbusier, the leading light of Modernist architecture and creator of the Modernist residential high-rise, proclaimed prophetically in 1937, "I propose one single building for all nations and climates." The gray concrete monoliths inspired by his designs, which once would have been livable nowhere, are now lived in everywhere—with air conditioners, elevators, and showers kept running on massive energy and water flows.[59]

About half of the energy use in building construction and operation is devoted to producing an artificial indoor climate—heating, cooling, ventilation, and lighting—so the potential savings from designs that co-opt natural forces to do the same thing can be quite large. Thus climate-sensitive design may be the biggest way to reduce the environmental impact of new buildings.[60]

The methods, though, vary from structure to structure. Small buildings have more heat-leaking or -absorbing exterior surface relative to the amount of indoor space, which makes insulation a priority. In contrast, large buildings lose less heat when it is cold outside, and gain less when it is warm—but special techniques are called for to make the best use of the sunlight they capture for illumination. In addition, climate-sensitive design must, by definition, adapt to regional variations in the availability of resources like sunlight, wind, and rain.[61]

Like warm-blooded animals, all buildings—but particularly small ones—need effective skins to control internal temperatures throughout the day and the year. The oil shocks of the seventies catapulted this simple idea into public awareness in many countries, leading millions to add insulation to their walls and roofs and to weather-strip their windows. Partly as a result, the amount of artificial heating required for each square meter in an average home in the United States fell by more than 40 percent between 1973 and 1990; in Denmark, the reduction was more than 46 percent.[62]

Over the coming 20 years, even greater gains are possible in countries of the former Eastern bloc, where central economic planning previously ignored energy wastefulness. As these countries begin to use markets to price energy, they will create the economic incentive for improving efficiency that the West experienced two decades ago. In the short term, however, a lack of capital may prevent building owners from investing in efficiency unless Western institutions provide assistance. Demonstrating the potential for cooperation, the Swedish government recently funded a retrofit project in Tallinn, Estonia, that employed simple measures, including insulation and weather stripping, to reduce energy use in apartment buildings by 30 percent. Given the greater reductions already achieved in the United States and Denmark, and the great inefficiency of the building stock in the former Eastern bloc, the full potential for cost-effective improvement could easily exceed 50 percent.[63]

Meanwhile, some builders in the West have discovered that designing from the ground up for air tightness and thermal integrity can make new buildings even more efficient than upgraded old ones. During the past decade, more than 100,000 "superinsulated" homes with extra-thick insulation and careful construction to block air leaks have been built in Scandinavia and North America. So hermetic are these homes that the warmth from people, lights, and appliances goes a long way toward heating them.[64]

Of course, if airtight buildings are poorly designed, pollutants such as cigarette smoke, molds, VOCs, and radon can build up inside. Most researchers, however, believe that efficient structures can have healthy indoor air if they are fitted with ventilation systems that run year round. Even more important is attacking the sources of pollutants by selecting low-VOC materials, and sealing basement walls and floors to block radon seepage from the earth.[65]

Climate-sensitive design may be the biggest way to reduce the environmental impact of new buildings.

New technologies have particularly revolutionized windows, which are often the leakiest parts of buildings. A variety of advanced windows have become popular in North America. All come with two or three panes of glass separated by insulating layers of air or argon gas, and some even have thin metallic coatings that allow visible light through but block invisible infrared and ultraviolet radiation (which transmit only heat). The residential market share in the United States for these types of windows rose from 1 percent in 1985 to 39 percent in 1993 as some manufacturers moved their whole product lines to the new technologies. (See Figure 3.)[66]

This market shift now saves homeowners some $5 billion in heating and cooling bills each year—an impressive figure, but one that falls far short of the $22 billion in additional savings that is technologically feasible. Not only do advanced windows insulate up to six times as well as single-pane glazings, but since they still admit sunlight, they can capture more energy than they lose. Winners of Canada's Advanced Houses competition demonstrated that, even at high latitudes, homes that combine improved windows with superinsulation can meet at least half of their already-low heat needs by letting in winter sunlight.[67]

Europe has developed an equally effective but radically different approach to energy-efficient fenestration. There, homes and offices commonly sport shading systems outside their windows. Some work like venetian blinds, while others unfold or roll shut. Such exterior shading systems can block light to pre-

FIGURE 3

U.S. Residential Market for Various Glazing Technologies, 1974–93

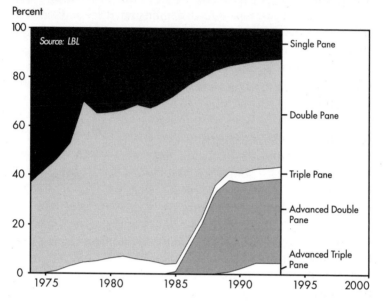

Percent

Source: LBL

- Single Pane
- Double Pane
- Triple Pane
- Advanced Double Pane
- Advanced Triple Pane

vent overheating as needed. They can also insulate if pulled down on a cold night, and provide a measure of physical security as well.[68]

Buildings in temperate regions can get most of their heat from the sun. The technique for doing this dates back at least to the ancient Greeks, who, when confronted with a fuelwood supply crisis, literally turned to the sun for heat, orienting many of their buildings with large south-facing openings. These structures captured solar rays in winter, when the sun rode low in the sky—exactly when it was most needed. But during the summer, when the sun climbed high overhead, they blocked direct light. Likewise, ancient Romans, Chinese, Japanese, and Anasazi Indians of North America all paid careful attention to the location of the sun when planning their buildings.[69]

Contrary to popular perception, compact cities can exploit solar energy effectively. Passive solar residences can be built as densely as 35 to 50 dwellings to the hectare, yet a typical U.S. residential suburb is zoned for no more than 10 homes per hectare. The German Weimar Republic, squeezed for cash in the 1920s by billion-dollar war reparations, demonstrated the suitability of this approach when it built whole neighborhoods of homes that captured the sun's heat to save on energy expenditures. And modern-day Berlin is again experimenting with the idea in one project, a group of six-story apartment buildings with south-facing solaria.[70]

Some architects today also exploit what is known as daylighting, which uses skylights, atria, and other techniques to reduce the need for electric light. Studies show, not surprisingly, that natural light is the best type of light for the human eye, and that proximity to windows improves well-being. For schools and offices that are occupied primarily during the day, daylighting is particularly cost-effective, cutting peak electricity demand. At the same time, it improves lighting quality, aesthetics, and worker satisfaction. More complicated technologies, such as window-mounted mirrors, or "light shelves," can spread natural light and its benefits deep into a building.[71]

Just as reservoirs set aside water from heavy rains to tide cities over during dry times, another basic climatic strategy—storage—allows buildings to use natural resources more efficiently. For example, most buildings could be made to channel rain into toilet tanks instead of sloughing it off into storm drains, reducing consumption of potable water. And buildings, like the Davis House, can also store heat. People in many traditional cultures in dry climates, including the Ladakhi in the Himalaya and the Pueblo in what is now New Mexico, make their homes with massive adobe walls that absorb heat and the sun's rays during the day—when there is too much—and return it to the air at night—when there is too little—moderating temperatures around the clock. Builders can also use water tanks, clay tiles, or concrete blocks to do the same thing.[72]

In the hotter climates more common to developing countries, and in bigger buildings, the design challenge is not making the most of solar heat, but getting rid of it. Along with proper solar orientation, deep eaves and recessed windows can shield building interiors from the sun. Air movement is also important. Perched atop most traditional homes in Hyderabad, Pakistan, are air scoops that capture the prevailing wind and draw it down to circulate through each story. Traditional homes in the southern United States included large porches to give people a place to sit in the long summer evenings, protected from the sun but exposed to cooling breezes.[73]

Field studies suggest that planting trees around some buildings provides enough shade to cut cooling needs by up to 30 percent. The energy reduction derived from using light-colored roofing and cladding materials that reflect sunlight instead of absorbing it can reach 40 percent. And insulation and insulating windows do just as good a job of keeping heat out as they do keeping it in. As with the Davis house, cutting cooling load in a new structure saves on energy bills and on the capital cost of air conditioning equipment, producing an immediate net savings. For society as a whole, it can also help forestall another major cost: that of new power plant construction. One study suggests that, in Thailand, the $10 million investment needed to build a small advanced window factory would, from the first year's production alone, save enough electricity to eliminate the need to commission a $1.5 billion power plant.[74]

Perhaps no one has experimented with integrating climate-sensitive techniques into twentieth century building forms as much as Malaysian architect Ken Yeang. In designing skyscrapers for the hot, humid tropics, Yeang first orients his buildings so as to minimize solar gain and to capture prevailing breezes to assist natural ventilation. He then incorporates sun shading—often with vegetation-filled balconies or courtyards high above the ground—to further block direct sunlight. In addition, wind ducts draw fresh air into the inner areas of the building. Yeang hopes to provide an alternative model for architects, particularly those in developing countries, who are accustomed to simply copying the standard Chicago or New York high-rise.[75]

Taken as a whole, climate-sensitive design using available technologies in the United States could cut heating and cooling energy use by 70 percent in residential buildings, and total energy use by 60 percent in commercial buildings, according to scientists at the government's National Renewable Energy Laboratory (NREL) in Golden, Colorado.[76]

Existing examples suggest that these estimates are quite conservative. By using solar orientation, thermal storage, and superinsulation in their new home, two NREL researchers cut heating bills by 97.5 percent compared to their neighbors'. Not far away, the Rocky Mountain Institute's headquarters gains 99 percent of its heat from the sun and the people and appliances inside, despite a frigid mountain climate.[77]

Undoubtedly, the potential savings in the rest of the world are comparable to the ones NREL estimates for the United States. Since buildings in operation use one-third of the world's energy, costing some $400 billion a year, cutting that use by half or more with climate-oriented design could eventually reduce the pollution from global energy use by a sixth and save roughly $200 billion annually. What makes this all the more astounding is how easy it would **Planting trees around buildings can cut cooling needs up to 30 percent.** be technically, mostly a matter of moving windows and judiciously adding insulation or ventilation.[78]

As Winston Churchill once said, "We shape our dwellings, and afterwards our dwellings shape our lives." One value of climate-sensitive buildings is the way they make visible once again the dependence of people on nature. Climate-sensitive buildings can help make for climate-sensitive people.[79]

William McDonough, dean of the University of Virginia architecture school in Charlottesville, has designed a German daycare center that illustrates this point well. A skylight would run the length of the building, providing much of its heat and light, and allowing children to follow the sun as it crossed the sky each day. If the building overheated or nap time came, the children themselves would be able to pull shutters across the windows, in effect "putting the building to sleep." Thus the building's design is educational in a fundamental way. It would teach young people to appreciate what most designers have forgotten: the relationship between the built environment and the natural one.[80]

Machines for Living

Le Corbusier once summed up the conventional twentieth century vision of buildings succinctly when he called the house "a machine for living." Indeed, wiring, piping, and duct work course invisibly through the walls and floors of modern buildings, feeding appliances with energy, water, and air, and carrying away waste. No matter how much environmental damage it might avert, eliminating the machinery that provides amenities such as indoor plumbing and refrigeration would make buildings less functional. The challenge, then, is to make machines in buildings as efficient as possible.[81]

Unfortunately, the inherent complexity of machinery can lead to waste. Because mechanical systems are hard to see and understand, they often malfunction and go unrepaired. For example, nearly all new homes in the United States feature forced-air climate control, and typically up to 30 percent of the heating and cooling energy they use escapes unnoticed through leaky or uninsulated duct work. Stories are also common of long-unobserved malfunctions in commercial buildings that have contributed to high utility bills and to sick building syndrome. Thus the first challenge in making modern building systems efficient is making them easier to monitor, maintain, and fix.[82]

Moreover, well-designed technologies can dramatically improve building efficiency. The history of lighting has demonstrated this repeatedly. Incandescent lamps are more than 10 times as efficient as oil lamps, provide better light, and are more convenient. Modern compact fluorescent lamps (CFLs) are four times as efficient again, last even longer, and produce light of comparable quality. As a result, CFL sales have risen spectacularly in recent years, capturing 15 percent of the market from incandescents. In Japan, where electricity costs 13¢ a kilowatt-hour, net savings average $55 per bulb; not surprisingly, CFLs now fill 80 percent of the country's home fixtures. Standard long-tube fluorescents are also becoming more efficient, better for the eyes, and quieter, thanks to electronic ballasts, reflectively coated fixtures, and switches that automatically sense when people leave a room. "Task lights," such as simple desk lamps, can reduce energy use as well, putting light only where people need it.[83]

Likewise, household appliances and fixtures such as furnaces, toilets, and air conditioners have all become more efficient in recent years, more than covering their sometimes higher up-front costs. Electricity use in new U.S. refrigerators, for instance, fell by 60 percent between 1972 and 1993, thanks to better insulation, more efficient electric motors, and other modest improvements; a new model introduced in 1994 uses even 30 percent less electricity and no ozone-depleting chlorofluorocarbons. New toilets and showerheads can cut water use in half. Technologies likely to emerge from the laboratory during the next decade could cut total appliance energy and water use in the United States by at least 25 percent. (See Table 3.) Moreover, in buildings with air conditioning, every energy efficiency gain carries an extra bonus, since the less power a machine consumes, the less heat it generates, saving both energy and capital costs for cooling equipment.[84]

KBI, a large nonprofit housing association in Denmark, recently exploited improvements like these when it upgraded a neighborhood of apartment buildings. In adding another floor to the top of each three-story building, it also fitted all the apartments with efficient new appliances, so that total energy and water use stayed the same. (The 33 percent gain in floor space also entailed much less material use than new construction would have, and at two-thirds the cost per square meter.)[85]

While using energy and water more efficiently, buildings can also tap renewable, on-site supplies. Solar water heaters offer a simple method of producing hot water on site without burning fossil fuels or splitting atoms. Popular in California, Florida, and Australia at the turn of the century before fossil fuels became inexpensive, they experienced a revival after World War II and again after the Arab oil embargo of 1973. More than 900,000 solar units in Israel heat 83 percent of the country's domestic hot water, and in Japan, some 4.5 million units were in place by 1992. Residents of Botswana's capital, Gaborone, have installed more than 3,000 solar water heaters, displacing nearly 15 percent of the country's residential elec-

TABLE 3

Resource Use of New U.S. Appliances and Prototypes in Early 1990s, Compared with Average Appliances Sold in 1985

Technology	Average For Sale, Early 1990s	Best For Sale, Early 1990s	Prototypes
		(percent reduction[1])	
Electricity			
Refrigerator	18	35	55–82
Central Air Conditioner	5	45	52–59
Water Heater	6	66	71–77
Gas			
Furnace	8	23	23
Water Heater	8	20	36
Cooking Range	20	40	60
Water			
Toilet	54	100[2]	100[2]
Showerhead	38	50	63

[1]Indicates reduction in resource use compared with average of appliances that were for sale in 1985. [2]Composting and incinerating toilets use no water.

Sources: Howard S. Geller, "Energy-Efficient Appliances: Performance Issues and Policy Options," *IEEE Technology and Society Magazine,* March 1986; Mark D. Levine et al., "Electricity End-Use Efficiency: Experience with Technologies, Markets, and Policies Throughout the World," American Council for an Energy-Efficient Economy (ACEEE), Washington, D.C., 1992; John Morrill, ACEEE, Washington, D.C., private communication, May 21, 1993; Steven Nadel et al., "Emerging Technologies in the Residential & Commercial Sectors," ACEEE, Washington, D.C., 1993; Amy Vickers, "Water-Use Efficiency Standards for Plumbing Fixtures: Benefits of National Legislation," *Journal of the American Water Works Association,* May 1990; Amy Vickers, Amy Vickers & Associates, Boston, Mass., private communication, September 6, 1994.

tricity demand. Some 30,000 are in use in Colombia, 17,000 in Kenya, and 10,000 in the Netherlands.[86]

Buildings can also supply their own electricity—and do so without producing carbon dioxide or nuclear waste. Solar photovoltaic cells produce electricity directly from sunlight with no moving parts. Driven by improvements in technology, as well as the adoption of mass production techniques, the cost of electricity from solar cells has fallen more than 90 percent since 1980.[87]

As solar cell prices continue to decline, integrating them right into a building's facade or roof, rather than attaching separate solar panels, may soon become common. Flachglas, a large German glass manufacturer, has integrated solar cells into a semitransparent window glazing that provides filtered light while generating electricity, installing several grid-connected prototypes in commercial buildings in seven German cities. Meanwhile, companies in Japan, Switzerland, and the United States are testing new types of solar cells that also function as roofing shingles or tiles. In western Germany, rooftop photovoltaics on existing buildings could generate up to 25 percent of the region's electricity. Even in the cloudy United Kingdom, the potential in recladding building facades with solar cells is estimated to equal half the country's electricity supply.[88]

True to the tenets of integrative design, as architects push the technological limits of efficiency, pieces of building machinery that were once separate become connected. For instance, buildings can save water by rerouting treated "gray water" from sinks and baths to toilets—a measure that is growing in attractiveness for water-short regions.

More than 900,000 solar units in Israel heat 83 percent of the country's domestic hot water.

Airtight homes provide another example of efficiency through interconnection. To keep the indoor air fresh, they usually depend on mechanical ventilators; the problem with this equipment is that in winter, the incoming air is much colder than the outgoing air, wasting heat (and vice versa in summer). Some designers get around this by adding heat exchangers to the ventilators, which extract warmth from one air flow and move it to the other. Using this arrangement, winning entries in Canada's Advanced Houses contest require so little artificial warmth that the heater has become a small add-on to the ventilation system.[89]

As buildings become less environmentally destructive, they will indeed become integrated, well-tuned machines for living. Someday, for instance, many buildings may "cogenerate" their own electricity, burning natural gas or renewably generated hydrogen, so that the waste heat can be used on site—rather than be dumped into the air or water by a large power plant far away. Such a basement generator might consist of a small turbine or an efficient battery-like device called a fuel cell. A single system would then provide heating and hot water, as well as power for lights, appliances, ventilation, and cooling.[90]

Better for Living, Better for Working

The pragmatic pitch for resource-efficient building has always been that it saves money on utility bills. Yet as significant as these savings can be, there is a growing realization within the building industry that the full benefits of ecological design are many times greater. Such design can greatly enhance the qualities buildings ought to have, making them not only more affordable, but also more pleasant and healthy—productive to work in, and desirable to live in. Translated into dollar terms, the increases in worker productivity or home value alone can far exceed the utility bill savings. (See Table 4.)

Residential developers apparently caught on to this fact sooner than commercial ones. Not far from the Davis House is a pioneering late-1970s development called Village Homes. The developer, Michael Corbett, narrowed the streets to discourage driving in favor of walking and biking, and ran them east-west so that he could orient houses toward the sun, which provides most of the space and water heating. Lots are small, and homes cluster around common areas. In summer, trees shade the streets and houses. Village Homes gives its residents a sense of connection to nature and neighbors that has pushed home values up 12 percent compared to those in similar, nearby developments.[91]

More recently, others have begun to follow the lead of Village Homes. For instance, by late 1993, two developers, Nick Martin

TABLE 4
Benefits of Ecological Design, Selected Examples

Building, Year Completed	Measures Undertaken, Cost	Outcome (annual gains)
Reno Post Office, Reno, Nevada, 1986	Lighting upgrade and lowered ceiling height to improve lighting quality and efficiency at cost of $300,000	$50,000 in energy and maintenance; $500,000 in productivity
Pennsylvania Power and Light, Allentown, Pennsylvania, early 1980s	Lighting upgrade and reorientation of fixtures in drafting engineers' office at cost of $8,362	73 percent drop in energy and maintenance; $42,240 (13 percent) gain in productivity
Internationale Nederlanden Bank, Amsterdam, 1987[1]	New building used energy efficient design for lighting, heating, and elimination of air conditioning; operable windows; thermal storage; cogeneration system; and avoidance of toxic materials at added cost of $700,000	$2.4 million in energy; $1 million (15 percent) drop in absenteeism
Village Homes, Davis, California, 1975–1981	220-home subdivision designed to capture 50–75 percent of heat from sun, incentives for non-motorized transportation, natural drainage, and edible landscape	12 percent premium in average home value
Lockheed Building 157, Sunnyvale, California, 1983	New building used daylighting, efficient lights, and an open layout to encourage worker interaction at added cost of $2 million	$500,000 in energy; $2 million (15 percent) drop in absenteeism; 15 percent gain in productivity
Esperanza del Sol, Dallas, Texas, 1994	New residential construction of low-income, energy efficient and solar-oriented houses at cost of $13 added annually to mortgage payments	$450 in energy

[1]The Internationale Nederlanden Bank was known as Nederlandsche Middenstandsbank when the new headquarters was inaugurated.

Sources: Joseph J. Romm, *Lean and Clean Management: How to Boost Profits and Productivity by Reducing Pollution* (New York: Kodansha International, 1994); Kim Hamilton, "Village Homes," *In Context,* Late Spring 1993; Cynthia Martin, Coldwell Banker/Doug Arnold Real Estate, Davis, Calif., private communication, February 7, 1995; Burke Miller Thayer, "Esperanza del Sol: Sustainable, Affordable Housing," *Solar Today,* May/June 1994.

in Nottingham, England, and John Clark in Fredericksburg, Virginia, were both planning environmentally oriented developments and, before they had even broken ground, they found themselves with long waiting lists of interested buyers.[92]

A low-income housing project in Dallas called Esperanza del Sol is using energy efficiency to make housing more affordable. Like the Davis House, these buildings contain less wood and more insulation in the walls, large insulating windows on the south side, airtight construction, and other features, to reduce energy use for cooling by 30 percent and for heating by 60 percent. Because the houses need smaller air conditioners and air ducts, the net additional cost is only $150 per home, amounting to $13 per year in extra mortgage payments. The developer, Barbara Harwood of BBH Enterprises in Carrollton, Texas, predicts that energy bills will be $450 lower a year. Daylighting and superior ventilation also give the houses a note of beauty and comfort often lacking in ones costing twice as much.[93]

Researchers at the Catholic University of Chile in Santiago have focussed on the idea that flexible designs can result in homes appropriate for low-income families. Their system uses three basic elements that builders can combine into a variety of floor plans. Load-bearing columns with T- or L-shaped footprints form the corners of the rooms, and are connected with lighter wall panels. The pieces are easy to build with, inexpensive because they can be mass produced, and durable because they are concrete. And since the walls can easily be knocked out, families can gradually add new rooms, allowing the homes to grow over time and develop individuality.[94]

Perhaps the most ambitious example of environmentally conscious home design is the government-initiated Ecolonia housing project in the Netherlands. Built in 1992, each of Ecolonia's 101 homes had to satisfy stringent requirements on materials choices, energy efficiency, and indoor air quality, while undertaking additional measures in one of these areas. Some save rainwater for the toilet tanks; some have roofs of sod; most use recycled concrete, solar water heaters, and integrated heating and ventilation systems. In all the units, the architects paid particular attention to the choice of paints and materials to avoid indoor pollution.[95]

Because they were pioneering, experimental buildings, Ecolonia's homes cost roughly 10 percent extra, but buyers have snatched them up anyway, and are apparently pleased. For many people, the added benefits already outweighed the extra costs, and as the new approach to construction has caught on elsewhere in the Netherlands, the premium has fallen to 5 percent. "Within a decade, I reckon this will be normal [Dutch] construction," says one of the project's architects.[96]

Such a shift may occur even faster in Sweden, where one of the country's three largest home builders, John Mattsson, announced in 1994 that it would only construct healthy and highly efficient buildings from then on. Small Swedish builders have already put up roughly 300 such houses in the last 10 years. Although the earliest homes cost slightly more than conventional ones, the differential has now all but disappeared—even before subtracting the typical $1,600 annual energy savings.[97]

What ecological homes do for families, ecological offices do for businesses. Since efficient buildings need less mechanical heating and cooling as well as smaller air ducts, designers can reduce the hidden area between floors. In a highly efficient structure, this would save enough height to insert an extra story for every four, giving a developer 25 percent more space to rent or sell for a modest additional outlay, thereby making the project more profitable.[98]

Such buildings' greatest benefits may accrue not to the developers but to the occupants. Features like natural light, fresh air, and user-adjustable task lights make offices more pleasant and give employees more control over their environment. Happier with their workplace, they show up for work more often, and are more productive. In a typical U.S. office, salaries are so high that if productivity rises just 2 percent, it is worth more to a company than eliminating utility bills entirely.[99]

A recent review of eight buildings conducted by Joseph Romm of the U.S. Department of Energy and William Browning of the Rocky Mountain Institute identified a gain of 6–16 percent in the productivity of people who work in such buildings. For example, the fall in absenteeism at the ING Bank's new headquarters saved $1 million annually, substantially adding to energy savings—a pattern found in other innovative buildings. A 1983 facility for the Lockheed Corporation in Sunnyvale, California, used daylighting, automatic light dimmers, and other features to cut lighting bills by $500,000 per year, which covered the $2 million additional cost in four years. In addition, worker productivity reportedly jumped 15 percent, a bonus worth at least an additional $2 million a year, and probably much more.[100]

The impressive successes of these housing developments and office buildings show that environmental design can give developers an edge in what is often an unforgivingly competitive business. Once enough innovators pursue the integrative design approach, they will reach the critical mass necessary to trigger a transformation of the building industry. This will vindicate the approach, which emphasizes thinking about all aspects of a building's operation and can yield unanticipated benefits. As Village Homes creator, Michael Corbett, once observed of his work, "You know you're on the right track when you notice that your solution for one problem accidentally solved several other problems."[101]

Blueprint for Better Buildings

Worldwide, the building industry is beginning to recognize the shortcomings of its products, and to discover that there are readily available, cost-effective remedies. In fact, some pioneering developers are beginning to use and market these alternatives, an encouraging sign of change.

As much as it is changing, however, the industry still has far to go. Many of the problems it contributes to, from the risk of climate change to species and habitat destruction, are worsening at an accelerating rate. Given the magnitude of these problems, institutions that can take the long view, such as governments, educators, and the lenders that broker most capital for construction, can all play a critical role. In fashioning a comprehensive strategy, they need to combine a number of tactics, including tightening building codes, taking steps to educate professionals and the public, and creating fiscal incentives that reward good building.

Governments have long taken an active role in the building sector through construction codes to ensure that structures are resistant to earthquakes and fires. In Kobe, Japan, code revisions in 1971 and 1981 appear to have kept the death toll from the 1995 earthquake from rising any higher than 5,000, since few of the newer structures collapsed. Energy and water-use codes and appliance standards, which, for example, set minimum levels of insulation or maximum levels of water use, have also appeared in the last 20 years. In 1978, California adopted a new energy code that had saved $11.4 billion on energy expenditures by 1995, and is predicted to save another $43 billion by 2011. Some countries are extending building codes to protect indoor air quality. The European Union has developed uniform standards, and the U.S. government released preliminary regulations in 1994.[102]

Codes are crucial for developing countries, where building booms are driving up energy and water use. In 1989, Mexico adopted strict limits on water use in new plumbing fixtures as part of a program to reduce per capita water withdrawals in its water-poor capital by a sixth. And in 1994, Thailand, facing rapid growth in electricity consumption, adopted an energy code much like those in the United States.[103]

Enforcing codes, though, can be difficult, especially in developing countries, where governance is often weak. Since 1986, for instance, China has required new apartments to be 30 percent more efficient, and in 1993 it raised the level to 50 percent, but builders and local governments alike have ignored the code. Even in affluent countries—for example in Dade County, Florida, where Hurricane Andrew came ashore—codes are sometimes flaunted.[104]

Codes and standards, though important, do little to educate consumers or encourage builders to innovate beyond the norm, as is essential to the development of a vibrant, environmentally conscious building market in the long run. As New York architect Randolph Croxton notes, any developer who proudly claims that a structure "meets every code" may actually be making a confession: "If I built this building any worse, it would be against the law."[105]

One of the first steps toward a more fundamental transformation of the industry will be changing the values it holds as a culture, and the process that transmits them, namely, education. Architectural and engineering students, in particular, need to learn to care about how their buildings work after they are built, not just how they look on the drawing board beforehand. Post-occupancy evaluations of buildings, now done infrequently, need to become routine, so that designers receive feedback from building users and managers on the ways that their buildings affect people, and on how well they survive the passage of time. In addition, professional schools need to teach students about integrative, interdisciplinary design processes.[106]

Government and industry groups should also work to educate practicing professionals. The 1994 success of the environmental code of practice published in the United Kingdom suggests that in many countries they will find a receptive audience. In 1991, the American Institute of Architects began publishing an Environmental Resource Guide that covers energy, land use, and detailed assessments of the impacts of various materials, providing a useful tool to practitioners.[107]

Equally needed are efforts to educate the general public in order to stimulate consumer interest in environmental techniques and technologies and accelerate their dissemination and commercialization. At relatively little cost, high-profile projects like Ecolonia and the ING Bank headquarters can have large impacts, as can design competitions. The ING Bank is now the best-known building in the Netherlands, according to national surveys.[108]

Other countries as well are launching demonstration projects. On Earth Day 1993, President Bill Clinton announced the "Greening of the White House." By late 1994, some 50 measures had been initiated, including lighting and water fixture upgrades, with additional steps planned for the next 20 years. The demonstration value of a project like this—seen by 1.5 million visitors a year—is far greater than the immediate savings. Operating on the same principle, the Thai government announced in 1994 that it would put up a 25-story office building in Bangkok that uses only 20 percent of the energy of a standard building, thereby stimulating the use of passive design and advanced cooling technologies suited to tropical climates.[109]

Design contests intended to spark innovative solutions are also catching on around the world. The Canadian government has continued in the Advanced Houses tradition with a new competition for commercial buildings, and three winning designs are now being built. All will use low-impact materials and half the normal amount of energy. Governments have also run competitions in New Zealand, as well as in France, where 700 homes based on the winning design will be built.[110]

To push these new approaches further into the building marketplace, some governments and industry groups are using voluntary rating systems. The U.K. government initiated a rating program in 1991, awarding points for features that go beyond code requirements—by saving energy and water, increasing recycled materials use, lowering toxic materials use, or reducing local environmental impacts. By mid-1994, more than 25 percent of new commercial buildings were being rated, and the government had developed programs to rate existing commercial buildings and new homes. Some British real estate agents now use high environmental marks to promote their properties.[111]

The Thai government has commissioned an office building that will use 80 percent less energy than others.

Rating is catching on in other parts of Europe, with governments in France, Norway, and Spain closest to adopting their own systems. An industry-developed system for commercial facilities in British Columbia has spread to Ontario, and the government of Quebec has expressed interest in it. In the United States, the nonprofit Green Building Council expects to launch a rating scheme for commercial buildings in 1995. In countries with great climatic and geographical variety, such as the United States, rating systems should vary regionally, leaving room for local organizations to contribute. Austin, Texas, for example, already has a 10-year-old Green Builder program that, among other things, rates residential construction.[112]

A variant of rating is labelling—instead of getting graded, buildings simply pass or fail. In 1980, the Canadian government

began a voluntary certification system for energy-efficient houses. Although only 8,000 homes have undergone certification to
earn the "R-2000" imprimatur, the program has improved mainstream construction by commercializing and popularizing a
number of efficient technologies. And in 1994, the government added requirements on materials selection and indoor
toxics sources. In general, creators of rating and labelling systems
should be prepared to revise them as scientific understanding of
environmental problems improves and as new materials and
technologies reach the market.[113]

Support from financial institutions is also crucial to trans-
forming the building market. Most building owners buy their
properties on credit, so banks, insurers, and other lenders can
substantially influence what gets built. Usually lenders see innovative projects as risky and subject them to costly delays or
higher interest rates. A few banks, however, have realized that
high-quality, resource-efficient buildings hold their value better
and leave owners more money to repay loans, making them a
better risk. The Bank of Montreal, for example, cuts the interest rate on loans for R-2000 homes by one-quarter point.[114]

In the United States, "energy-efficient mortgages," which
reduce the income requirements on energy-efficient homes,
have been available through federal and state lending agencies
as well as private banks for more than a decade. Most U.S.
financial institutions, though, resist making more significant
changes. Sweden, in marked contrast, has subsidized loans for
energy-efficient homes for 20 years, which partly explains how
the Swedish building industry has reached such high standards
of efficiency. And in early 1995, Sweden's largest housing bank,
Hypoteksbanken, announced that, having sustained massive
losses from sick building syndrome, it plans to lend money *only*
for "green" buildings. For the same reason, insurance companies
could also begin to reward healthier buildings with lower liability
coverage premiums.[115]

Multilateral financial institutions, led by the World Bank,
should help developing countries improve their fast-growing
building stock. The bank lends hundreds of millions of dollars
each year to East European and developing countries for pro-

jects relating to housing; it has recently emphasized improving the institutional framework of housing ministries, mainly to remove barriers to housing construction, and providing adequate water and sanitation services in burgeoning cities. But in its effort to ensure basic services, the bank has only recently started to consider specific measures to improve energy and water efficiency.[116]

Electric, natural gas, and water utilities can also play a role in ensuring efficient use of natural resources, in order to reduce expenditures on fuel and new power and water treatment plants. Some electric utilities give rebates to building owners for upgrading lights, appliances, and mechanical systems; but utilities can also be more proactive, targeting new building construction. One approach would be a "feebate" system that charges less-efficient structures for connections to energy and water utilities, and then uses this revenue to reward developers of more-efficient buildings. A few North American utilities are trying a similar method, helping home buyers with the down payments on new houses that are energy-efficient. New Brunswick Power tenders rebates for R-2000 homes, which is one reason these now account for 20 percent of the new houses built in that province—the highest rate in Canada. Until budget problems forced it to cut the program in 1995, Ontario Hydro offered rebates to design teams to cut energy use in new buildings, and provided additional money once savings were demonstrated. Water and energy utilities could also perform home inspections on a regular basis, checking for inefficient heaters and fixtures, and for leaky duct work and pipes as well.[117]

Sweden's largest housing bank has announced that it plans to lend money only for "green" buildings.

Just as utilities are reaching into the building process, governments can develop constructive partnerships with industry. One consequence of the fragmentation of the building business in most market economies is that architectural, engineering, and construction firms are too financially insecure to invest in research and development. Production of some components, such as windows and ceiling tiles, tends to be more concentrated, and here innovation continues at a steady pace. What can result, for example, is a home with advanced windows and leaky walls. Research needs to focus not only on new technologies but also on techniques for integrating them. The Swedish government spends more per capita on building research than most countries, another reason that homes there are so energy efficient.[118]

The U.S. Environmental Protection Agency has initiated an innovative set of voluntary programs to spur businesses to upgrade their facilities; it serves as a clearinghouse for information on new technologies, and provides tools for predicting the economics of their deployment in specific situations. The Green Lights program, which has more than 1,500 participants so far, will eventually reduce lighting energy use by about half in at least 5 percent of U.S. commercial floor space. The Energy Star Buildings program, inaugurated in 1994, carries the approach further, guiding building managers through a series of steps to improve air quality and cut energy use in existing structures an impressive 50 percent—all cost-effectively.[119]

Governments should also execute fundamental reforms of fiscal policy. Currently, most governments subsidize, rather than tax, pollution and resource extraction. The U.S. government gives the mining industry a $500 million annual tax break. The German government spends $4.9 billion a year subsidizing coal production at more than four times the world price; 36 percent of this coal is burned to provide heat and power to buildings. Subsidies for coal and hydropower development are also endemic in developing countries, where the average price of electricity—30 percent of which is used in buildings—fell from 7.0¢ in 1983 to just 4.9¢ only five years later.[120]

Governments need to gradually eliminate such subsidies and reform tax policy so that the true economic and social costs of depletion and pollution are passed on to those who benefit from it. If governments balance such increases with cuts in other taxes—on labor, for example—the net impact could be positive, providing simultaneous incentives for efficiency and employment. Governments might also channel a small percentage of the money into research and development, helping to accelerate the expansion of resource-efficient and healthy construction.[121]

Perhaps no country has moved as aggressively as the Netherlands in formulating a broad buildings program. The Dutch campaign is part of the country's 1989 National Environment Policy Plan (NEPP), updated in 1993, which addresses a wide spectrum of environmental issues. Among the goals of NEPP's sustainable housing initiative are: to improve energy efficiency by 25 percent in new and renovated buildings by the year 2000, to increase the reuse and recycling of construction and demolition waste from 60 percent in 1990 to 90 percent by 2000, to halve the use of toxic materials such as oil-based paints, and to eliminate the use of unsustainably produced tropical timber.[122]

Since 1990, the Dutch government has worked with architects, developers, contractors, building associations, and residents' associations to assemble an action plan to meet NEPP goals. Meanwhile, a nongovernmental organization, the Association for Integral Biological Architecture (VIBA), runs an information center that recently received government funding; there, some 100 companies sell "green" products certified by VIBA and provide information for builders and architects. The government has also backed high-profile demonstration projects, such as Ecolonia. In addition, the NEPP calls for taxes on virgin materials and on waste disposal to encourage recycling of building materials and more efficient use of fuels and water.[123]

Around the world, the challenge for policymakers, educators, and lenders resembles that facing builders. No single measure, whether tax reform or solar design, will fully rectify the fundamental problems with modern buildings. Rather, what is needed is a broad set of changes in the way everyone involved does business. Like an integrative design team, society can enter the process of change with a rough sense of the path it will take, but it needs a clear idea of its destination. The goal is one all can easily agree on: creating a secure home for humanity.

Notes

1. Figure of 90 percent from U.S. Environmental Protection Agency (EPA), Office of Air and Radiation (OAR), *Report to Congress on Indoor Air Quality, Volume II: Assessment and Control of Indoor Air Pollution* (Washington, D.C.: 1989). Where buildings are put can have as much impact on people's health and the environment—by increasing the need for motorized transportation—as how they are made and operated. For coverage of these issues, see Marcia D. Lowe, *Shaping Cities: The Environmental and Human Dimensions*, Worldwatch Paper 105 (Washington, D.C.: October 1991).

2. Quote from Marlise Simons, "Earth-Friendly Dutch Homes Use Sod and Science," *New York Times*, March 7, 1994.

3. Carbon dioxide concentration rise from H. Friedli et al., "Ice Core Record of the $^{13}C/^{12}C$ Ratio of Atmospheric CO_2 in the Past Two Centuries," *Nature*, November 20, 1986, and from Timothy Whorf, Scripps Institution of Oceanography, La Jolla, Calif., private communication, February 2, 1995; emissions share is a Worldwatch estimate, based on R.M. Rotty and G. Marland, "Production of CO_2 from Fossil-Fuel Burning" (electronic database) (Oak Ridge, Tenn.: Oak Ridge National Laboratory (ORNL), 1993), on G. Marland and T.A. Boden, "Global, Regional, and National CO_2 Emission Estimates from Fossil Fuel Burning, Cement Production, and Gas Flaring: 1950–1990" (electronic database) (Oak Ridge, Tenn.: ORNL, 1993), on Organisation for Economic Co-operation and Development (OECD), International Energy Agency (IEA), *Energy Balances of OECD Countries 1960–79* (Paris: 1991), on OECD, IEA, *World Energy Statistics and Balances 1971–1987* (Paris: 1989), and on OECD, IEA, *Energy Statistics and Balances of Non-OECD Countries 1991–1992* (Paris: 1994).

4. United Nations (UN), *Monthly Bulletin of Statistics* (New York: various issues); Worldwatch estimate, based on UN, *Long-range World Population Projections: Two Centuries of Population Growth, 1950–2150* (New York: 1992).

5. Figure 1 and South Korea are from UN, *Yearbook of Construction Statistics* (New York: various years), from UN, *Construction Statistics Yearbook* (New York: various years), and from UN, *Monthly Bulletin of Statistics*, op. cit. note 4.

6. Figure 2 is from Lee Schipper and Stephen Meyers, *Energy Efficiency and Human Activity: Past Trends and Future Prospects* (Cambridge, U.K.: Cambridge University Press, 1992), and from L. Schipper and C. Sheinbaum, "Recent Trends in Household Energy Use Efficiency in OECD Countries: Stagnation or Improvement," in *Proceedings of 1994 ACEEE Summer Study on Energy Efficiency in Buildings* (Washington, D.C.: American Council for an Energy-Efficient Economy (ACEEE), 1994); Alan Thein Durning, *Saving the Forests: What Will It Take?* Worldwatch Paper 117 (Washington, D.C.: Worldwatch Institute, December 1993); Ahluwalia is from Mitchell Owens, "Building Small...Thinking Big," *New York Times*, July 21, 1994.

7. Robert Ornstein and Paul Ehrlich, *New World New Mind* (New York: Simon and Schuster, Inc., 1989).

8. Charles Smiler, "Building Energy Efficiency in Commercial and Investment Real

Estate," unpublished manuscript, Real Estate Skills for the Environment, Montpelier, Vt., undated; Amory B. Lovins, "Energy-Efficient Buildings: Institutional Barriers and Opportunities," Strategic Issues Paper, E SOURCE, Inc., Boulder, Colo., 1992.

9. Lovins, op. cit. note 8.

10. U.S. Congress, Office of Technology Assessment (OTA), *Energy Efficiency Technologies for Central and Eastern Europe* (Washington, D.C.: U.S. Government Printing Office (GPO), 1993); Yu Joe Huang, "Potential for and Barriers to Building Energy Conservation in China," *Contemporary Policy Issues* (California State University, Long Beach), July 1990.

11. Ajay Shanker, "Constructing Sustainable Houses in Hurricane Prone Florida," in Charles J. Kibert, ed., *Supplemental Proceedings of the First International Conference of CIB TG 16* (Gainesville, Fla.: University of Florida, 1994); "fastest-growing" from Patricia Gober, "Americans on the Move," *Population Bulletin*, November 1993; "thousands" is a conservative Worldwatch estimate, based on Shanker, op. cit. this note, and on Rod Miner, Center for Resourceful Building Technology (CRBT), Missoula, Mont., private communications, February 6, 7, and 10, 1995.

12. Matthys Levy and Mario Salvadori, *Why Buildings Fall Down* (New York: W.W. Norton & Company, 1992); Michael Melkumian, World Bank, Yerevan, Armenia, private comminication, December 21, 1994.

13. Gober, op. cit. note 11; Stewart Brand, *How Buildings Learn: What Happens After They're Built* (New York: Viking Penguin, 1994).

14. Brand, op. cit. note 13.

15. Urs Morf, "Modernization and Nostalgia in Beijing," *Swiss Review of World Affairs*, September 1994.

16. OTA, *Building Energy Efficiency* (Washington, D.C.: GPO, 1992); description of U.S. and Swedish factories from Lee Schipper, Stephen Meyers, and Henry Kelly, *Coming in From the Cold: Energy-Wise Housing in Sweden* (Cabin John, Md.: Seven Locks Press, 1985).

17. John Picard, Environmental Enterprises, Inc., Marina del Rey, Calif., private communication, July 26, 1994.

18. The Internationale Nederlanden (ING) Bank was called the Nederlandsche Middenstandsbank (NMB Bank) until 1991; ING Bank, "Building with a Difference: ING Bank Head Office," Amsterdam, undated; Bill Holdsworth, "Organic Services," *Building Services*, March 1989.

19. Brenda Vale and Robert Vale, *Green Architecture: Design for an Energy-conscious Future* (Boston: Bulfinch Press, 1991); William Browning, "NMB Bank Headquarters," *Urban Land*, June 1992; Rob Vonk, ING Bank, Amsterdam, private communication, March 25, 1994.

20. Francis Duffy, "Measuring Building Performance," *Facilities*, May 1990, cited in Brand, op. cit. note 13.

21. Brand, op. cit. note 13.

22. Ibid.; Pei's preference for fixed uses from Carter Wiseman, *I.M. Pei: A Profile* *in American Architecture* (New York: Henry N. Abrams, Inc., 1990).

23. Brand, op. cit. note 13.

24. National Audubon Society and Croxton Collaborative, Architects, *Audubon House: Building the Environmentally Responsible, Energy-Efficient Office* (New York: John Wiley & Sons, 1994).

25. John Bennett, *International Construction Project Management: General Theory and Practice* (Oxford: Butterworth Heinemann, 1991).

26. Hideo Imamura, Shimizu Corporation, Tokyo, private communications, January 5, 1995, and February 1 and 2, 1995; Yasuyoshi Miyatake, Executive Vice President, Shimizu Corporation, "Sustainable Construction," speech given at the First International Conference of CIB TG 16, Tampa, Fla., November 7, 1994.

27. Germany from Varis Bokalders, Royal Institute of Technology, Stockholm, private communication, February 2, 1995; "Mainstream Architects Going Green," *Environmental Building News*, January/February 1995; S.P. Halliday, *Environmental Code of Practice for Buildings and Their Services* (Bracknell, U.K.: Building Services Research and Information Association (BSRIA), May 1994); Zoe Crawford, BSRIA, Bracknell, U.K., private communication, July 25, 1994.

28. Paul Oliver, *Dwellings: The House Across the World* (Austin: University of Texas Press, 1987); Frank Lloyd Wright cited in Victor Papenek, "'Dwellings: The House Across the World'" (book review), *Earthword*, No. 5, 1994.

29. James Howard Kunstler, *The Geography of Nowhere* (New York: Simon & Schuster, 1993).

30. Vale and Vale, op. cit. note 19.

31. Christopher Alexander, *The Timeless Way of Building* (New York: Oxford University Press, 1979).

32. Figure of 40 percent is an extrapolation from U.S. data based on global data from Donald G. Rogich, U.S. Department of the Interior (DOI), Bureau of Mines (BOM), "Changing Minerals and Material Use Patterns," presented at the Annual General Meeting of the Academia Europaea, Parma, Italy, June 23–25, 1994, and on Bill Kelleher, National Stone Association, Silver Spring, Md., private communication, July 13, 1994; John E. Young, *Mining the Earth*, Worldwatch Paper 109 (Washington, D.C.: Worldwatch Institute, July 1992).

33. Daniel Edelstein, DOI, BOM, Washington, D.C., private communication, July 19, 1994; share of copper recycled excludes "new scrap," waste metal that factories send back to the copper mills for recycling, it never having reached the consumer, and is based on ibid.; Young, op. cit. note 32; Sandra Kraemer, "Material Changes in the Building and Construction Industry: Piping Applications," in DOI, BOM, *The New Materials Society: Materials Shifts in the New Society, Vol. 3* (Washington, D.C.: 1991); Nadav Malin and Alex Wilson, "Should We Phase Out PVC?" *Environmental Building News*, January/February 1994; Keith Schneider, "E.P.A. Moves to Reduce Health Risks from Dioxin," *New York Times*, September 14, 1994; German position on PVC is from Lisa Finaldi, Greenpeace International, "PVC Debate Continues" (letter to the editor), *Environmental Building News*, November/December 1993; Rich Gilbert, American Public Health

Association, Washington, D.C., private communication, August 15, 1994.

34. Sandra Postel and John Ryan, "Reforming Forestry," in Lester R. Brown et al., *State of the World 1991* (New York: W.W. Norton & Company, 1991); Steve Loken, "Materials for a Sustainable Building Industry," *Solar Today*, November/December 1991; Schipper, Meyers, and Kelly, op. cit. note 16; International Tropical Timber Association (ITTO), *Annual Review and Assessment of the World Tropical Timber Situation, 1990–1991* (Yokohama, Japan: 1992); Nancy Chege, "Roundwood Production Unabated," in Lester R. Brown, Hal Kane, and David Malin Roodman, *Vital Signs 1994* (New York: W.W. Norton & Company, 1994); D.O. Hall, King's College, London, private communication and printout, March 7, 1994.

35. Durning, op. cit. note 6; ITTO, op. cit. note 34.

36. Figure of 150 tons is a Worldwatch estimate, based on the minerals content of a typical U.S. home, from Aldo F. Barsotti, DOI, BOM, Washington, D.C., private communication and printout, July 13, 1994; refuse figure is from Stephen D. Cosper, William H. Hallenbeck, and Gary R. Brenniman, "Construction and Demolition Waste: Generation, Regulation, Practices, Processing, and Policies," Office of Solid Waste Management, University of Illinois, Chicago, 1993; Jane Maynard, Bureau of the Census, Suitland, Md., private communication, August 5, 1994; U.S. waste rates from Robert Brickner, Gershman, Brickner & Bratton, Inc., Falls Church, Va., private communication, August 23, 1994; European waste rates from Erik K. Lauritzen and Niels Jørn Hahn, "Building Waste—Generation and Recycling," *International Solid Wastes & Public Cleansing Association Yearbook 1991–1992* (London: Associated Publishing Group plc, undated), and from OECD, *OECD Environmental Data: Compendium 1993* (Paris: 1993).

37. Worldwatch estimate, based on National Audubon Society and Croxton Collaborative, op. cit. note 24, on U.S. Department of Energy (DOE), Energy Information Administration (EIA), *Annual Energy Review 1993* (Washington, D.C.: GPO, 1994), and on Kelleher, op. cit. note 32.

38. Worldwatch estimates, based on OECD, *Energy Balances of OECD Countries 1960–79*, op. cit. note 3, on OECD, *World Energy Statistics and Balances 1971–1987*, op. cit. note 3, and on OECD, *Energy Statistics and Balances of Non-OECD Countries 1991–1992*, op. cit. note 3, with biomass estimates for developing countries from Hall, op. cit. note 34, and from OTA, *Energy in Developing Countries* (Washington, D.C.: GPO, 1991).

39. Worldwatch estimates, based on OECD, *Energy Statistics and Balances of Non-OECD Countries 1991–1992*, op. cit. note 3, on Hall, op. cit. note 34, on OTA, op. cit. note 38, on National Audubon Society and Croxton Collaborative, op. cit. note 24, on DOE, op. cit. note 37, and on Kelleher, op. cit. note 32.

40. Building's share of world water use is a Worldwatch estimate, based on Peter H. Gleick, "Water and Energy," in Peter H. Gleick, ed., *Water in Crisis: A Guide to the World's Fresh Water Resources* (New York: Oxford University Press, 1993), on OECD, *Energy Statistics and Balances of Non-OECD Countries 1991–1992*, op. cit. note 3, and on World Resources Institute, *World Resources 1994–95* (New York: Oxford University Press, 1994), and excludes water use for electricity production for manufacturing building materials; Sandra Postel, *Last Oasis: Facing Water Scarcity* (New York: W.W. Norton & Company, 1992).

41. Figure of 30 percent is from a 1984 World Health Organization committee report, cited in EPA, OAR, "Indoor Air Facts No. 4: Sick Building Syndrome," Washington, D.C., 1991; EPA, Office of Acid Deposition, Environmental Monitoring and Quality Assurance, *Indoor Air Quality in Public Buildings, Vol. I* (Washington, D.C.: 1988); National Audubon Society and Croxton Collaborative, op. cit. note 24.

42. David Mudarri, EPA, Indoor Air Division, private communication, Washington, D.C., July 15, 1994; William Fisk, Lawrence Berkeley Laboratory (LBL), Berkeley, Calif., private communication, July 15, 1994.

43. Nadav Malin, "Steel or Wood Framing: Which Way Should We Go?" *Environmental Building News*, July/August 1994; Nigel Howard, Davis Langdon Consultancy, London, private communication and printout, September 20, 1994.

44. Embodied energy of wood from Hall, op. cit note 34; poor suitability of wood in humid climates from Nadav Malin, Environmental Building News, Brattleboro, Vt., private communication, August 5, 1994.

45. Michael Moquin, "Adobe, Rammed Earth, & Mud: Ancient Solutions for Future Sustainability," *Earthword*, No. 5, 1994; European history from Brenda Vale, University of Nottingham, U.K., private communication, July 29, 1994; Marina Trappeniers, Craterre, Grenoble, France, private communication, September 6, 1994.

46. Jean Dethier, "A Back-to-the-earth Approach to Housing," *UNESCO Courier*, March 1985; Paul McHenry, University of New Mexico, Albuquerque, N.M., private communication, December 9, 1994.

47. Lynne Bayless, "Strawbale & Steel," *Earthword*, No. 5, 1994; David Bainbridge, "Plastered Straw Bale Construction: Super Energy Efficient and Economical," The Canelo Project, Canelo, Ariz., 1992; David Eisenberg, The Development Center for Appropriate Technology, Tucson, Ariz., private communication, February 22, 1995.

48. Rooney Massara, Stramit International, Suffolk, U.K., private communication, January 31, 1995; Michael Langley, Terra Verde, Austin, Tex., private communication, August 3, 1994; ecological cost is a Worldwatch estimate, based on ibid., on Ed Ayres, Worldwatch Insitute, Washington, D.C., private communication, February 13, 1995, and on Howard, op. cit. note 43.

49. National Audubon Society and Croxton Collaborative, op. cit. note 24; Steve Loken, CRBT, Missoula, Mont., private communication, August 10, 1994; Clint Good, Clint Good Architects, Lincoln, Va., private communication, February 6, 1995; Bokalders, op. cit. note 27.

50. Loken, op. cit. note 49; Miner, op. cit. note 11; Carl Weinberg, Weinberg and Associates, Walnut Creek, Calif., private communication, April 23, 1993.

51. Howard, op. cit. note 43; OTA, *Green Products by Design: Choices for a Cleaner Environment* (Washington, D.C.: GPO, 1992); Scott Gilliland, HDR Engineering,

Omaha, Neb. (subsidiary of Bouygues, S.A., Paris), private communication, December 1, 1994.

52. Cosper, Hallenbeck, and Brenniman, op. cit. note 36; Tim Mayo, "Canada's Advanced Houses Program," in A.H. Fanney et al., eds., *U.S. Green Buildings Conference—1994* (Washington, D.C.: GPO, 1994); "Nova Scotia's Advanced House: The EnviroHome," Nova Scotia Department of Natural Resources, Halifax, undated.

53. Vale and Vale, op. cit. note 19; Mark Wachle, Recycled Materials Company, Colorado Springs, Colo., private communication, August 11, 1994; Sydney from "Bulletin Board," *NCS Bulletin* (IUCN Pakistan), Karachi, September 1992.

54. Carole Douglis, "Making Houses Out of Trash," *World Watch*, November/December 1993; Robert Noble, Gridcore Systems International, Carlsbad, Calif., private communications, July 8 and 29, 1994.

55. Amory Lovins, "Hot-Climate House Predicted to Need No Air Conditioner, Cost Less to Build," Tech Memo, E SOURCE, Inc., Boulder, Colo., November 1993.

56. Ibid.

57. Ibid.

58. Ibid.; Lance Elberling, Pacific Gas and Electric Company, San Ramon, Calif., private communication, January 30, 1995.

59. Richard G. Stein, *Architecture and Energy: Conserving Energy Through Rational Design* (New York: Anchor Press, 1977); Stanley Schuler, *The Cape Cod House* (Exton, Pa.: Schiffer Publishing, Ltd., 1982); Brent C. Brolin, *The Failure of Modern Architecture* (New York: Van Nostrand Reinhold Company, 1976).

60. Worldwatch estimate, based on U.S. data from DOE, op. cit. note 37, from DOE, EIA, *Monthly Energy Review September 1993* (Washington, D.C.: GPO, 1993), from DOE, EIA, *Annual Energy Outlook 1994* (Washington, D.C.: GPO, 1994), and from Mohammad Adra, DOE, EIA, Washington, D.C., private communication, March 7, 1994, and on global data from OECD, *Energy Statistics and Balances of Non-OECD Countries 1991–1992*, op. cit. note 3, from Hall, op. cit. note 34, and from OTA, op. cit. note 38.

61. Steven Ternoey et al., *The Design of Energy-Responsive Commercial Buildings* (New York: John Wiley & Sons, 1985).

62. Residential energy use statistics are based on energy use per degree day per square meter of home area, and are from Schipper and Meyers, op. cit. note 6, and from Schipper and Sheinbaum, op. cit. note 6.

63. "NUTEK's Programme for an Environmentally Adapted Energy System in the Baltic Region and Eastern Europe," Swedish National Board for Industrial and Technical Development (NUTEK), Stockholm, November 1994; Hans Nilsson, NUTEK, Stockholm, private communication, February 2, 1995.

64. David Olivier, "The House that Came in from the Cold," *New Scientist*, March 9, 1991.

65. *Health and Environment Digest* (Freshwater Foundation, Wayzata, Minn.), May 1993; James Woods, Virginia Polytechnic Institute and State University, Blacksburg, Va., private communication, August 9, 1994.

66. Figure 3 considers windows to be "advanced" if they include low-E coating or argon fill, and is based on K. Frost, D. Arasteh, and J. Eto, "Savings from Energy Efficient Windows: Current and Future Savings from New Fenestration Technologies in the Residential Market," LBL, Berkeley, Calif., April 1993, and on Karl Frost, LBL, Berkeley, Calif., private communication, January 31, 1995.

67. Frost, Arasteh, and Eto, op. cit. note 66; Joan Gregerson et al., *Space Heating Technology Atlas* (Boulder, Colo.: E SOURCE, Inc., 1993); Mayo, op. cit. note 52.

68. Donald R. Johnsen, Hella U.S.A., Grafton, N.H., private communication, November 22, 1994. The reasons for the divergence between North American and European approaches seems to be deeply cultural, and may include greater willingness on the part of Europeans to accept devices that require occasional manual operation, and to accept, because of their longer historical perspective, the somewhat longer financial payback periods for external shading systems compared to advanced windows.

69. Ken Butti and John Perlin, *Golden Thread: 2500 Years of Solar Architecture and Technology* (Palo Alto, Calif.: Cheshire Books, 1980).

70. Susan E. Owens, "Land Use Planning for Energy Efficiency," in J.B. Cullingworth, ed., *Energy, Land, and Public Policy* (New Brunswick, N.J.: Transaction Publishers, 1990); Michael B. Brough, "Density and Dimensional Regulations, Article XII," *A Unified Development Ordinance* (Washington, D.C.: American Planning Association, Planners Press, 1985); Butti and Perlin, op. cit. note 69; Berlin from Paul Okamoto and Eric Saijo, "Ecological Site Design for Affordable Housing," *The Urban Ecologist*, No. 3, 1994.

71. Lindsay Audin et al., *Lighting Technology Atlas* (Boulder, Colo.: E SOURCE, Inc., 1994); Alicia Ravetto, "Daylighting Schools in North Carolina," *Solar Today*, March/April 1994; Audin et al., op. cit. this note.

72. Butti and Perlin, op. cit. note 69; Helena Norberg-Hodge, *Ancient Futures: Learning from Ladakh* (San Francisco: Sierra Club Books, 1991).

73. Ken Yeang, *Bioclimatic Skyscrapers* (London: Artemis, 1994); Stein, op. cit. note 59.

74. LBL and Sacramento Municipal Utility District, "Peak Power and Cooling Energy Savings of Shade Trees and White Surfaces: Year 2," LBL, Berkeley, Calif., April 27, 1993; Karen L. George, "Highly Reflective Roof Surfaces Reduce Cooling Energy Use and Peak Demand," Tech Update, E SOURCE, Inc., Boulder, Colo., December 1993; first year's production is for the 30-year service life of the windows, and is from Ashok Gadgil, Arthur H. Rosenfeld, and Lynn Price, "Making the Market Right for Environmentally Sound Energy-Efficient Technologies: U.S. Buildings Sector Successes that Might Work in Developing Countries and Eastern Europe," presented to ESETT '91: International Symposium on Environmentally Sound Energy Technologies and their Transfer to Developing Countries and European Economies in Transition, Milan, October 21–25, 1991.

75. Yeang, op. cit. note 73; Cathleen McGuigan, Stanley Holmes, and Maggie Malone, "Towers Rise in the East," *Newsweek*, July 18, 1994.

76. Doug Balcomb, National Renewable Energy Laboratory (NREL), Golden, Colo., private communication and printout, June 2, 1993.

77. Cecile M. Liboef and Craig Christensen, "The Minimum Energy House," *Solar Today*, January/February 1991; Vale and Vale, op. cit. note 19.

78. Figure of $400 billion is a conservative Worldwatch estimate, based on OECD, *Energy Balances of OECD Countries 1960–79*, op. cit. note 3, on OECD, *World Energy Statistics and Balances 1971–1987*, op. cit. note 3, and on OECD, *Energy Statistics and Balances of Non-OECD Countries 1991–1992*, op. cit. note 3, and on OECD, IEA, *Energy Prices and Taxes, Third Quarter, 1992* (Paris: 1992).

79. Churchill quote is from Brand, op. cit. note 13.

80. Michael Wagner, "Tuning in Children," *Interiors*, March 1993.

81. Kunstler, op. cit. note 29.

82. David A. Jump and Mark Modera, "Energy Impacts of Attic Duct Retrofits in Sacramento Houses," in *Proceedings of 1994 ACEEE Summer Study*, op. cit. note 6; Woods, op. cit. note 65; stories of waste in commercial buildings from Frank Kensill, Institute for Human Development, Philadelphia, Pa., private communication, October 1991, and from Lee Eng Lock, Supersymmetry Services Pte. Ltd., Singapore, private communication, August 11, 1994.

83. Robert van der Plas and A.B. de Graaff, "A Comparison of Lamps for Domestic Lighting in Developing Countries," World Bank, Washington, 1988; M.D. Levine et al., "Electricity End-Use Efficiency: Experience with Technologies, Markets, and Policies Throughout the World," ACEEE, Washington, D.C., 1992; market share is a Worldwatch estimate, taking into account that compact fluorescents last 10 times as long as incandescents, based on Nils Borg, NUTEK, Stockholm, Sweden, private communication, February 10, 1995; $55 figure is a Worldwatch estimate of the net present value of the payback from replacing a 75-watt, 1,000-hour incandescent bulb with a 19-watt, 10,000-hour CFL, using a 3 percent annual real rate of return on five-year savings, a price of 75¢ for incandescent bulbs, and a residential electricity price for Japan from OECD, op. cit. note 78; 80 percent figure is from Leslie Lamarre, "Shedding Light on the Compact Fluorescent," *EPRI Journal*, March 1993; Audin et al., op. cit. note 71.

84. Brooke Stauffer, Association of Home Appliance Manufacturers, Washington, D.C., private communication, December 9, 1993; David Goldstein, Natural Resources Defense Council, San Francisco, Calif., private communication, June 20, 1993; Steven Nadel et al., "Emerging Technologies in the Residential & Commercial Sectors," ACEEE, Washington, D.C., 1993; Amy Vickers, "The Energy Policy Act: Assessing Its Impact on Utilities," *Journal of the American Water Works Association*, August 1993.

85. Erik Toxværd Nielsen, Toftegård, Herlev, Denmark, private communication, April 20, 1994.

86. Butti and Perlin, op. cit. note 69; Josef Nowarski, Division of Research and Development, Ministry of Energy and Infrastructure, Jerusalem, private com-

munication, January 19, 1994; Eddie Bet Hazavdi, Energy Conservation Division, Ministry of Energy and Infrastructure, Jerusalem, private communication and printout, January 26, 1994; Solar System Development Association, "The Status of Solar Energy Systems in Japan," Tokyo, 1993; Botswana from Chris Neme, Memorandum to Mark Levine, LBL, Berkeley, Calif., March 28, 1992; Mario Calderón and Paolo Lugari, Centro Las Gaviotas, Bogota, Colombia, private communication, April 13, 1992; Kenya from Christopher Hurst, "Establishing New Markets for Mature Energy Equipment in Developing Countries; Experience with Windmills, Hydro-Powered Mills and Solar Water Heaters," *World Development*, Vol. 18, No. 4, 1990; E.H. Lysen, Netherlands Agency for Energy and the Environment (NOVEM), "Solar Energy in the Netherlands," presented to the IEA-SHCP National Programs Workshops, Sydney, Australia, May 4, 1993.

87. Cost of solar cells is from Paul Maycock, Photovoltaic Energy Systems, Inc., Casanova, Va., private communications and printouts, May 8, 1992, and December 20, 1993; for an expanded discussion on solar cells, see Christopher Flavin and Nicholas Lenssen, *Power Surge: Guide to the Coming Energy Revolution* (New York: W.W. Norton & Company, 1994).

88. Joachim Brenemann, "Energy Active Façades: Technology and Possibilities of Photovoltaic Integration into Buildings," Flachglas Solartechnik, Köln, Germany, undated; "Reference List," Flagsol, Flachglas Solartechnik Gmbh, Köln, Germany, January 14, 1994; Steven Strong, "An Overview of Worldwide Development Activity in Building-Integrated Photovoltaics," Solar Design Associates, Harvard, Mass., undated; western Germany from W.H. Bloss et al., "Grid-Connected Solar Houses," in *Proceedings of the 10th EC Photovoltaics Solar Energy Conference* (Dordrecht: Kluwer Academic Publishing, 1991); R. Hill, N.M. Pearsall, and P. Claiden, *The Potential Generating Capacity of PV-Clad Buildings in the UK*, Vol. 1 (London: Department for Trade and Industry, 1992); OECD, IEA, *Energy Policies of IEA Countries: 1991 Review* (Paris: 1992); examples assume a 25 percent capacity factor for photovoltaics.

89. Mayo, op. cit. note 52.

90. Keith G. Davidson and Gerald W. Braun, "Thinking Small: Onsite Power Generation May Soon Be Big," *Public Utilities Fortnightly*, July 1, 1993.

91. Kim Hamilton, "Village Homes," *In Context*, Late Spring 1993; John Tillman Lyle, *Regenerative Design for Sustainable Development* (New York: John Wiley & Sons, 1994); Cynthia Martin, Coldwell Banker/Doug Arnold Real Estate, Davis, Calif., private communication, February 7, 1995.

92. Brenda Vale and Robert Vale, University of Nottingham, U.K., private communication, September 15, 1994; John Clark, John A. Clark Co., Washington, D.C., private communication, July 12, 1994.

93. Burke Miller Thayer, "Esperanza del Sol: Sustainable, Affordable Housing," *Solar Today*, May/June 1994.

94. Guillermo Thenoux and Luis F. Alacón, "Reducing Waste and Site Construction Impacts in Housing Projects," in Charles J. Kibert, ed., *Proceedings of the First International Conference of CIB TG 16* (Gainesville, Fla.: University of Florida, 1994).

95. NOVEM, "Ecolonia: Demonstration Project for Energy-Saving and Environmentally-aware Building and Living," Sittard, The Netherlands, undated.

96. Simons, op. cit. note 2; Gonda van Hal, Association for Integral Biological Architecture, Den Bosch, The Netherlands, private communication, July 28, 1994.

97. Bokalders, op. cit. note 27.

98. Amory B. Lovins, "Designing Buildings for Greater Profit," presentation at the National Association of Home Builders, Washington, D.C., March 2, 1994.

99. Joseph J. Romm, *Lean and Clean Management: How to Boost Profits and Productivity by Reducing Pollution* (New York: Kodansha International, 1994).

100. Joseph J. Romm and William D. Browning, "Greening the Building and the Bottom Line: Increasing Productivity Through Energy-Efficient Design," in *Proceedings of 1994 ACEEE Summer Study*, op. cit. note 6.

101. Hamilton, op. cit. note 91.

102. Dennis Normile, "Quake Builds Strong Case for Codes," *Science*, Jaunary 27, 1995; savings for California energy code are net of incremental energy investments, assume no improvement would have take place without the codes in place, and are from Robert Schlichting, California Energy Commission, Sacramento, Calif., private communication, February 9, 1995; James Woods, Virginia Polytechnic Institute and State University, Blacksburg, Va., private communication, August 25, 1994; U.S. Department of Labor, Occupational Safety and Health Administration, *Federal Register*, April 5, 1994; Woods, op. cit. this note.

103. Mexico from Postel, op. cit. note 40; Soontorn Boonyatikarn, Chulalongkorn University, Bangkok, private communication, August 26, 1994.

104. Siwei Lang and Yu Joe Huang, "Energy Conservation Standard for Space Heating in Chinese Urban Residential Buildings," *Energy—The International Journal*, August 1993; Yu Joe Huang, LBL, Berkeley, Calif., private communication, September 26, 1994; Shanker, op. cit. note 11.

105. Randolph R. Croxton, "Foreword," in National Audubon Society and Croxton Collaborative, op. cit. note 24.

106. Infrequency of post-occupancy evaluations from Brand, op. cit. note 13.

107. Crawford, op. cit. note 27; Harry T. Gordon, "The American Institute of Architects Committee on the Environment," in Fanney et al., op. cit. note 52.

108. Vonk, op. cit. note 19.

109. The White House, Office on Environmental Policy (OEP), "The Greening of the White House: Phase I Action Plan," Washington, D.C., March 11, 1994; Brian Johnson, The White House, OEP, Washington, D.C., private communication, August 18, 1994; Boonyatikarn, op. cit. note 103.

110. "C-2000 Advanced Commercial Buildings Program," program literature, undated, Natural Resources Canada, Ottawa, Ont.; John B. Storey, "Eco-house: An Environment, User, Context Friendly Home," in Kibert, op. cit. note 94; Gilles Olive, Bureau d'Etudes Gilles Olive, Paris, private communication, November 23, 1994.

111. Building Research Establishment (BRE), "BREEAM/New Offices," Version 1/93, Garston, Watford, U.K.; John Doggart, ECD Partnership, London, private communication, July 28, 1994; BRE, "BREEAM/New Homes," Version 3/91, Garston, Watford, U.K.; BRE, "BREEAM/Existing Offices," Version 4/93, Garston, Watford, U.K.; Michael Scholand, "Buildings for the Future," *World Watch*, November/December 1993.

112. Doggart, op. cit. note 111; Roger Baldwin, BRE, Garston, Watford, U.K., private communication, August 15, 1994; Raymond Cole, University of British Columbia, "Building Environmental Performance Assessment Criteria (BEPAC)," in Fanney et al., op. cit. note 52; Craig Crawford, Green Workplace, Government of the Province of Ontario, Toronto, Ont., private communication, February 8, 1995; Tom Cohn, executive director, U.S. Green Building Council, Bethesda, Md., private communication, February 7, 1995; W. Lawrence Doxsey, "The City of Austin Green Builder Program," in Fanney et al., op. cit. note 52.

113. Gary Sharp, Post Harvest Developments, Inc., Ottawa, private communication, July 20, 1994.

114. Ibid.

115. Barbara Farhar and Jan Eckert, "Energy-Efficient Mortgages and Home Energy Rating Systems: A Report on the Nation's Progress," NREL, Golden, Colo., September 1993; Schipper, Meyers, and Kelly, op. cit. note 16; Lee Schipper, LBL, Berkeley, Calif., private communication, April 1, 1994; Bokalders, op. cit. note 27.

116. The World Bank, *Annual Report* (Washington, D.C.: various years).

117. Amory B. Lovins, "Negawatts for Development," Energy Efficiency Roundtable, World Bank, Washington, D.C., September 14, 1994; Gary Sharp, Post Harvest Developments, Inc., Ottawa, private communication, March 31, 1994; "R-2000 Reaches New Milestone," *R-2000 News Communiqué*, Energy, Mines, and Resources Canada, Ottawa, February 1993; David Grafstein, Ontario Hydro, Toronto, private communication, February 10, 1995.

118. Industry fragmentation from United Nations Economic Commission for Europe, *Annual Bulletin of Housing and Building Statistics for Europe* (New York: 1992); Schipper, Meyers, and Kelly, op. cit. note 16.

119. Robert Kwartin, EPA, OAR, Washington, D.C., private communication, September 13, 1994; EPA, OAR, "Introducing...The Energy Star Buildings Program," Washington, D.C., November 1993; Chris O'Brien, EPA, OAR, Washington, D.C., private communication, March 9, 1994.

120. U.S. Congress, Subcommittee on Natural Resources, Committee on Oversight and Investigations, "Taking from the Taxpayer: Public Subsidies for Natural Resource Development," April 1994; Judy Dempsey, "Decision Time Looms for German Energy," *Financial Times*, February 9, 1995; Worldwatch estimate, based on OECD, *Energy Balances of OECD Countries 1991–1992*, op. cit. note 3; Worldwatch estimate, based on OECD, *Energy Balances of Non-OECD Countries 1991–1992*, op. cit. note 3; World Bank, Energy Development Division, *Review of Electricity Tariffs in Developing Countries During the 1980's*, Industry and Energy Department Working Paper, Energy Series Paper No. 32 (Washington, D.C.: 1990).

121. Ernst U. Von Weizsäcker and Jochen Jesinghaus, *Ecological Tax Reform: A Policy Proposal for Sustainable Development* (London: Zed Books, 1992).

122. Ministry of Housing, Spatial Planning and the Environment (VROM), *The Netherlands' National Environmental Policy Plan 2* (The Hague: 1994); VROM, "Working with the Construction Sector," Environmental Policy in Action No. 2, The Hague, March 1994.

123. Van Hal, op. cit. note 96; VROM, "Working with the Construction Sector," op. cit. note 122; NOVEM, op. cit. note 95; VROM, *National Environmental Policy Plan*, op. cit. note 122.

PUBLICATION ORDER FORM

No. of
Copies

_____ 57. **Nuclear Power: The Market Test** by Christopher Flavin.
_____ 58. **Air Pollution, Acid Rain, and the Future of Forests** by Sandra Postel.
_____ 60. **Soil Erosion: Quiet Crisis in the World Economy** by Lester R. Brown and
Edward C. Wolf.
_____ 61. **Electricity's Future: The Shift to Efficiency and Small-Scale Power**
by Christopher Flavin.
_____ 63. **Energy Productivity: Key to Environmental Protection and Economic Progress**
by William U. Chandler.
_____ 65. **Reversing Africa's Decline** by Lester R. Brown and Edward C. Wolf.
_____ 66. **World Oil: Coping With the Dangers of Success** by Christopher Flavin.
_____ 67. **Conserving Water: The Untapped Alternative** by Sandra Postel.
_____ 68. **Banishing Tobacco** by William U. Chandler.
_____ 69. **Decommissioning: Nuclear Power's Missing Link** by Cynthia Pollock.
_____ 70. **Electricity For A Developing World: New Directions** by Christopher Flavin.
_____ 71. **Altering the Earth's Chemistry: Assessing the Risks** by Sandra Postel.
_____ 75. **Reassessing Nuclear Power: The Fallout From Chernobyl** by Christopher Flavin.
_____ 77. **The Future of Urbanization: Facing the Ecological and Economic Constraints**
by Lester R. Brown and Jodi L. Jacobson.
_____ 78. **On the Brink of Extinction: Conserving The Diversity of Life** by Edward C. Wolf.
_____ 79. **Defusing the Toxics Threat: Controlling Pesticides and Industrial Waste**
by Sandra Postel.
_____ 80. **Planning the Global Family** by Jodi L. Jacobson.
_____ 81. **Renewable Energy: Today's Contribution, Tomorrow's Promise** by
Cynthia Pollock Shea.
_____ 82. **Building on Success: The Age of Energy Efficiency** by Christopher Flavin
and Alan B. Durning.
_____ 83. **Reforesting the Earth** by Sandra Postel and Lori Heise.
_____ 84. **Rethinking the Role of the Automobile** by Michael Renner.
_____ 86. **Environmental Refugees: A Yardstick of Habitability** by Jodi L. Jacobson.
_____ 88. **Action at the Grassroots: Fighting Poverty and Environmental Decline**
by Alan B. Durning.
_____ 89. **National Security: The Economic and Environmental Dimensions** by Michael Renner.
_____ 90. **The Bicycle: Vehicle for a Small Planet** by Marcia D. Lowe.
_____ 91. **Slowing Global Warming: A Worldwide Strategy** by Christopher Flavin
_____ 92. **Poverty and the Environment: Reversing the Downward Spiral** by Alan B. Durning.
_____ 93. **Water for Agriculture: Facing the Limits** by Sandra Postel.
_____ 94. **Clearing the Air: A Global Agenda** by Hilary F. French.
_____ 95. **Apartheid's Environmental Toll** by Alan B. Durning.
_____ 96. **Swords Into Plowshares: Converting to a Peace Economy** by Michael Renner.
_____ 97. **The Global Politics of Abortion** by Jodi L. Jacobson.
_____ 98. **Alternatives to the Automobile: Transport for Livable Cities** by Marcia D. Lowe.
_____ 99. **Green Revolutions: Environmental Reconstruction in Eastern Europe and the
Soviet Union** by Hilary F. French.
_____100. **Beyond the Petroleum Age: Designing a Solar Economy** by Christopher Flavin
and Nicholas Lenssen.
_____101. **Discarding the Throwaway Society** by John E. Young.
_____102. **Women's Reproductive Health: The Silent Emergency** by Jodi L. Jacobson.
_____103. **Taking Stock: Animal Farming and the Environment** by Alan B. Durning and
Holly B. Brough.
_____104. **Jobs in a Sustainable Economy** by Michael Renner.
_____105. **Shaping Cities: The Environmental and Human Dimensions** by Marcia D. Lowe.
_____106. **Nuclear Waste: The Problem That Won't Go Away** by Nicholas Lenssen.

☐ **Single Copy: $5.00**

☐ **Bulk Copies (any combination of titles)**
   ☐ 2–5: $4.00 ea.   ☐ 6–20: $3.00 ea.   ☐ 21 or more: $2.00 ea.
   Inquire for discounts on larger orders.

☐ **Membership in the Worldwatch Library: $30.00 (international airmail $45.00)**
   The paperback edition of our 250-page "annual physical of the planet,"
   State of the World, plus all Worldwatch Papers released during the calendar year.

☐ **Worldwatch Database Disk: $89**
Includes up-to-the-minute global agricultural, energy, economic, environmental, social,
and military indicators from all current Worldwatch publications.

Please check one: _____high-density IBM-compatible or   _____Macintosh
---

☐ **Subscription to *World Watch* Magazine: $20.00 (international airmail $35.00)**
   Stay abreast of global environmental trends and issues with our award-
   winning, eminently readable bimonthly magazine.

Please include $3 postage and handling for non-subscription orders.

Make check payable to Worldwatch Institute
1776 Massachusetts Avenue, N.W., Washington, D.C. 20036-1904 USA

Enclosed is my check for U.S. $_____

VISA ☐   MasterCard ☐   _____
                              Card Number   Expiration Date

---
name   **daytime phone #**

---
address

---
city   **state**   **zip/country**

Phone: (202) 452-1999   Fax: (202) 296-7365   E-Mail: wwpub@igc.apc.org   WWP

PUBLICATION ORDER FORM

No. of
Copies

_____ 57. **Nuclear Power: The Market Test** by Christopher Flavin.
_____ 58. **Air Pollution, Acid Rain, and the Future of Forests** by Sandra Postel.
_____ 60. **Soil Erosion: Quiet Crisis in the World Economy** by Lester R. Brown and Edward C. Wolf.
_____ 61. **Electricity's Future: The Shift to Efficiency and Small-Scale Power** by Christopher Flavin.
_____ 63. **Energy Productivity: Key to Environmental Protection and Economic Progress** by William U. Chandler.
_____ 65. **Reversing Africa's Decline** by Lester R. Brown and Edward C. Wolf.
_____ 66. **World Oil: Coping With the Dangers of Success** by Christopher Flavin.
_____ 67. **Conserving Water: The Untapped Alternative** by Sandra Postel.
_____ 68. **Banishing Tobacco** by William U. Chandler.
_____ 69. **Decommissioning: Nuclear Power's Missing Link** by Cynthia Pollock.
_____ 70. **Electricity For A Developing World: New Directions** by Christopher Flavin.
_____ 71. **Altering the Earth's Chemistry: Assessing the Risks** by Sandra Postel.
_____ 75. **Reassessing Nuclear Power: The Fallout From Chernobyl** by Christopher Flavin.
_____ 77. **The Future of Urbanization: Facing the Ecological and Economic Constraints** by Lester R. Brown and Jodi L. Jacobson.
_____ 78. **On the Brink of Extinction: Conserving The Diversity of Life** by Edward C. Wolf.
_____ 79. **Defusing the Toxics Threat: Controlling Pesticides and Industrial Waste** by Sandra Postel.
_____ 80. **Planning the Global Family** by Jodi L. Jacobson.
_____ 81. **Renewable Energy: Today's Contribution, Tomorrow's Promise** by Cynthia Pollock Shea.
_____ 82. **Building on Success: The Age of Energy Efficiency** by Christopher Flavin and Alan B. Durning.
_____ 83. **Reforesting the Earth** by Sandra Postel and Lori Heise.
_____ 84. **Rethinking the Role of the Automobile** by Michael Renner.
_____ 86. **Environmental Refugees: A Yardstick of Habitability** by Jodi L. Jacobson.
_____ 88. **Action at the Grassroots: Fighting Poverty and Environmental Decline** by Alan B. Durning.
_____ 89. **National Security: The Economic and Environmental Dimensions** by Michael Renner.
_____ 90. **The Bicycle: Vehicle for a Small Planet** by Marcia D. Lowe.
_____ 91. **Slowing Global Warming: A Worldwide Strategy** by Christopher Flavin
_____ 92. **Poverty and the Environment: Reversing the Downward Spiral** by Alan B. Durning.
_____ 93. **Water for Agriculture: Facing the Limits** by Sandra Postel.
_____ 94. **Clearing the Air: A Global Agenda** by Hilary F. French.
_____ 95. **Apartheid's Environmental Toll** by Alan B. Durning.
_____ 96. **Swords Into Plowshares: Converting to a Peace Economy** by Michael Renner.
_____ 97. **The Global Politics of Abortion** by Jodi L. Jacobson.
_____ 98. **Alternatives to the Automobile: Transport for Livable Cities** by Marcia D. Lowe.
_____ 99. **Green Revolutions: Environmental Reconstruction in Eastern Europe and the Soviet Union** by Hilary F. French.
_____100. **Beyond the Petroleum Age: Designing a Solar Economy** by Christopher Flavin and Nicholas Lenssen.
_____101. **Discarding the Throwaway Society** by John E. Young.
_____102. **Women's Reproductive Health: The Silent Emergency** by Jodi L. Jacobson.
_____103. **Taking Stock: Animal Farming and the Environment** by Alan B. Durning and Holly B. Brough.
_____104. **Jobs in a Sustainable Economy** by Michael Renner.
_____105. **Shaping Cities: The Environmental and Human Dimensions** by Marcia D. Lowe.
_____106. **Nuclear Waste: The Problem That Won't Go Away** by Nicholas Lenssen.

☐ **Single Copy: $5.00**

☐ **Bulk Copies (any combination of titles)**
    ☐ 2–5: $4.00 ea.   ☐ 6–20: $3.00 ea.   ☐ 21 or more: $2.00 ea.
Inquire for discounts on larger orders.

☐ **Membership in the Worldwatch Library: $30.00 (international airmail $45.00)**
The paperback edition of our 250-page "annual physical of the planet,"
State of the World, plus all Worldwatch Papers released during the calendar year.

☐ **Worldwatch Database Disk: $89**

Includes up-to-the-minute global agricultural, energy, economic, environmental, social,
and military indicators from all current Worldwatch publications.

Please check one: _____high-density IBM-compatible or _____Macintosh

---

☐ **Subscription to *World Watch* Magazine: $20.00 (international airmail $35.00)**
Stay abreast of global environmental trends and issues with our award-
winning, eminently readable bimonthly magazine.

Please include $3 postage and handling for non-subscription orders.

Make check payable to Worldwatch Institute
1776 Massachusetts Avenue, N.W., Washington, D.C. 20036-1904 USA

Enclosed is my check for U.S. $_____

VISA ☐   MasterCard ☐ _____
                         Card Number   Expiration Date

_____
name   **daytime phone #**

_____
address

_____
city   **state**   **zip/country**

Phone: (202) 452-1999   Fax: (202) 296-7365   E-Mail: wwpub@igc.apc.org   WWP